# Advanced Virtual Assistants - A Window to the Virtual Future

*Edited by Ali Soofastaei*

Published in London, United Kingdom

Advanced Virtual Assistants - A Window to the Virtual Future
http://dx.doi.org/10.5772/intechopen.1000365
Edited by Ali Soofastaei

Contributors
Adegboyega Adegboye, Ali Soofastaei, Hans Luyten, Hideyuki Namba, James Hutson, Daniel Plate, Jeff Stanley, Jinsoon Song



First published in London, United Kingdom, 2024 by IntechOpen
IntechOpen is the global imprint of INTECHOPEN LIMITED, registered in England and Wales, registration number: 11086078, 5 Princes Gate Court, London, SW7 2QJ, United Kingdom

British Library Cataloguing-in-Publication Data
A catalogue record for this book is available from the British Library

Additional hard and PDF copies can be obtained from orders@intechopen.com

Advanced Virtual Assistants - A Window to the Virtual Future
Edited by Ali Soofastaei
p. cm.
Print ISBN 978-1-83769-712-0
Online ISBN 978-1-83769-711-3
eBook (PDF) ISBN 978-1-83769-713-7

For EU product safety concerns:
IN TECH d.o.o., Prolaz Marije Krucifikse Kozulić 3, 51000 Rijeka, Croatia,
info@intechopen.com or visit our website at intechopen.com.

# Meet the editor

Dr. Ali Soofastaei is a global artificial intelligence (AI) projects leader, an international keynote speaker, and a professional author. He completed his Ph.D. and was a postdoctoral research fellow at the University of Queensland, Australia, in the field of AI applications in mining engineering, where he led a revolution in the use of deep learning and AI methods to increase energy efficiency, reduce operation and maintenance costs, and reduce greenhouse gas emissions in surface mines. As a scientific supervisor, he has provided practical guidance to undergraduate and postgraduate students in mechanical and mining engineering and information technology for many years. Dr. Soofastaei has more than 20 years of academic experience as an assistant professor and leader of global research activities. Results from his research and development projects have been published in international journals and keynote presentations. He has presented his practical achievements at conferences in the United States, Europe, Asia, and Australia. He has been involved in industrial research and development projects in several industries, including oil and gas (Royal Dutch Shell), steel (Danieli), and mining (BHP, Rio Tinto, Anglo American, Vale, and Newmont). His extensive practical experience in the industry has equipped him to work with complex industrial problems in highly technical and multidisciplinary teams.

# Contents

# Preface

The technological realm is transforming, with Advanced Virtual Assistants (AVAs) standing at the forefront, redefining the boundaries of Human–Computer Interaction (HCI). *Advanced Virtual Assistants – A Window to the Virtual Future* delves deep into this transformation, exploring the myriad dimensions of AVAs and their implications for various sectors. Each chapter in this book offers a unique perspective, addressing diverse aspects of and challenges posed by AVAs.

*Advanced Virtual Assistants – A Window to the Virtual Future* offers a panoramic view of AVAs' advancements, applications, and challenges in diverse fields. Through the lens of each chapter, readers are invited to explore the multifaceted world of virtual assistants and envision their transformative potential for the future. This book is a valuable resource for researchers, professionals, and anyone intrigued by the evolution of HCI in the digital age.

As we traverse the fascinating terrain of AVAs, this book acts as both a guide and a catalyst for critical thought. Each chapter is meticulously crafted to unravel the complexities and potential of AVAs, serving as stepping stones for future innovations and developments in the field.

Chapter 1 lays the groundwork by exploring how AVAs revolutionize HCI. It introduces the reader to the fundamental principles behind AVAs, providing a comprehensive overview of their evolution, current capabilities, and potential for transforming interactions in the digital age.

Chapter 2's in-depth exploration of the interplay between ICT and physical activity sheds light on how technology and well-being are intertwined, opening avenues for holistic approaches to health and fitness.

As we navigate the intricacies of human–AI collaboration in education, delineated in Chapter 3, we witness the metamorphosis of learning experiences and the enrichment of metacognitive capabilities, laying the foundation for a more enlightened society.

Further, exploring voice-based chatbots in public service in Chapter 4 unveils the multifaceted aspects of digital inclusion and public communication, setting the stage for a more inclusive and communicative future.

As we delve into the personality of virtual assistants through a self-presentation approach, as discussed in Chapter 5, we are led to ponder the essence of user experience and the myriad ways AVAs can be molded to be more relatable and engaging.

Chapter 6's elucidation of the fundamental concepts and algorithms in machine learning demystifies the backbone of AVAs, providing clarity and insight into the driving forces behind their intelligence and adaptability.

Reflecting on the historical perspective in Chapter 7, we gain a deeper understanding of the correlation between the proliferation of online chatting and reading literacy developments, thus appreciating digital communication's transformative impact on literacy skills.

Concluding with an exploration of the application of virtual assistants in modern mining, chapter 8 illuminates the diverse potentials of AVAs across different sectors, highlighting their role in enhancing operational efficiency, ensuring worker safety, and fostering innovation.

This comprehensive exploration across diverse chapters encapsulates the multifarious nature of AVAs, providing readers with a well-rounded understanding and appreciation of their transformative potential. By weaving together varied perspectives, insights, and applications, *Advanced Virtual Assistants – A Window to the Virtual Future* stands as a testament to HCI's boundless possibilities and promising future.

We hope readers, whether researchers, professionals, students, or enthusiasts, find this book a source of knowledge, inspiration, and insight into the rapidly evolving world of AVAs. As we stand on the cusp of a virtual future, this book invites us to envision the possibilities and actively participate in shaping the trajectory of HCI for the betterment of society.

**Dr. Ali Soofastaei**
CEO,
Innovative AI (IAI),
Brisbane, Australia

Chapter 1

# Introductory Chapter: Advanced Virtual Assistants – Transforming the Future of Human-Computer Interaction

*Ali Soofastaei*

## 1. Introduction

The rapid advancements in artificial intelligence (AI) and machine learning technologies have given rise to a new era of human-computer interaction. Advanced virtual assistants are at the forefront of this transformation, providing users with a seamless, efficient, and intelligent way of accomplishing tasks, accessing information, and interacting with digital devices. This chapter introduces advanced virtual assistants, their underlying technologies, applications, and their impact on various aspects of our lives. By the end of this book, readers will have a thorough understanding of advanced virtual assistants, their potential, and the exciting opportunities they present for the future [1].

### 1.1 What are advanced virtual assistants?

Advanced virtual assistants are AI-powered software agents designed to assist users in performing tasks, retrieving information, and providing personalized recommendations. They can understand natural language, process information, and respond intelligently to user requests. These virtual assistants go beyond the capabilities of traditional rule-based systems by leveraging machine learning, natural language processing (NLP), and other cutting-edge AI technologies. This allows them to learn, adapt, and improve over time, providing a more intuitive and human-like experience for the user.

### 1.2 Evolution of virtual assistants

The concept of virtual assistants can be traced back to early computer programs that aimed to simulate human conversation, such as ELIZA and SHRDLU. However, as we know them today, the first generation of virtual assistants emerged as simple voice-activated assistants, like Apple's Siri and Google Assistant. These assistants relied on predefined rules and basic algorithms to perform limited tasks and answer queries.

The second generation of virtual assistants incorporated more sophisticated AI technologies, such as machine learning and NLP, enabling them to better understand user intent, context, and preferences. This led to the development of advanced virtual assistants, such as OpenAI's GPT-3, capable of carrying out complex tasks, holding more in-depth conversations, and providing personalized experiences.

### 1.3 Key technologies powering advanced virtual assistants

Advanced virtual assistants are built upon various AI technologies that enable them to understand, reason, and respond to user requests. Some of the key technologies include:

- *Machine Learning:* Advanced virtual assistants utilize machine learning algorithms to learn from user interactions and adapt their responses accordingly. This enables them to improve and provide more accurate and relevant information.

- *Natural Language Processing (NLP):* NLP allows virtual assistants to understand and process human language, enabling them to communicate effectively with users and perform tasks using natural language commands.

- *Speech Recognition:* Advanced virtual assistants rely on speech recognition technology to convert spoken language into text, allowing users to interact with them using voice commands.

- *Sentiment Analysis:* Sentiment analysis enables virtual assistants to understand the emotional context of user requests and provide more empathetic and personalized responses.

### 1.4 Applications of advanced virtual assistants

Advanced virtual assistants have a wide range of applications across various industries, including:

- *Personal Assistance:* Advanced virtual assistants can help users manage their schedules, set reminders, answer questions, and provide recommendations based on their preferences.

- *Customer Support:* Virtual assistants can provide customer support, answer queries, troubleshoot issues, and guide users through various processes.

- *Healthcare:* Advanced virtual assistants can assist medical professionals with scheduling appointments, managing patient records, and providing medical advice based on patient data.

- *Education:* Virtual assistants can be used as personalized learning assistants, helping students with study materials, answering questions, and providing feedback on their progress.

- *E-commerce:* Virtual assistants can assist users in finding products, answering questions about products and services, and processing transactions.

### 1.5 The future of advanced virtual assistants

As AI technologies continue to evolve, the capabilities of advanced virtual assistants will expand and become increasingly integrated into various aspects of our lives. Some of the potential developments in the future of advanced virtual assistants include:

- *More human-like interactions:* With advancements in natural language understanding and generation, advanced virtual assistants can engage in more meaningful, context-aware conversations with users, providing a more human-like interaction experience.

- *Enhanced personalization:* As virtual assistants continue learning and adapting to individual user preferences, they can provide highly personalized experiences tailored to each user's needs and interests.

- *Cross-platform integration:* Advanced virtual assistants will become increasingly interconnected with various digital devices and platforms, allowing users to seamlessly manage and interact with multiple services and applications using a single virtual assistant.

- *Proactive assistance:* In the future, advanced virtual assistants can anticipate user needs and provide proactive assistance, offering relevant information, suggestions, and reminders without explicit user requests.

- *Collaboration with other AI systems:* Advanced virtual assistants will collaborate with other AI systems to perform complex tasks that require the coordination of multiple specialized AI agents, leading to a more efficient and effective problem-solving process.

## 2. Technical foundations of advanced virtual assistants

To fully appreciate the capabilities and potential of advanced virtual assistants, it is essential to understand the technical foundations that underpin these systems. Therefore, we will explore the key AI technologies and techniques that enable virtual assistants to understand, reason, and respond to user requests intelligently [2].

### 2.1 Machine learning

Machine learning is a subset of AI that allows systems to learn and improve from experience without being explicitly programmed. Advanced virtual assistants leverage machine learning algorithms to process and analyze vast amounts of data, enabling them to recognize patterns, make predictions, and generate responses based on their input.

There are several types of machine learning techniques used in virtual assistants, including:

- *Supervised Learning:* The algorithm is trained on a labeled dataset of input-output pairs. The algorithm learns the relationship between the input and output, allowing it to make predictions on new, unseen data.

- *Unsupervised Learning:* Unsupervised learning algorithms work with unlabeled data and learn to identify patterns, relationships, or structures within the data without prior knowledge of the desired output.

- *Reinforcement Learning:* Reinforcement learning algorithms learn by interacting with their environment and receiving feedback through rewards or penalties. This allows the algorithm to discover the optimal actions in a given situation to maximize compensation.

## 2.2 Natural language processing (NLP)

Natural Language Processing (NLP) is an area of AI focusing on the interaction between computers and human language. NLP enables advanced virtual assistants to understand, interpret, and generate human language, allowing them to communicate effectively with users and perform tasks using natural language commands.

Some of the key NLP techniques used in virtual assistants include:

- *Tokenization:* Tokenization breaks text into individual words, phrases, or symbols, allowing the algorithm to analyze and process the text more effectively.

- *Part-of-Speech (POS) Tagging:* POS tagging involves assigning a grammatical category (e.g., noun, verb, adjective) to each token in the text, which helps the algorithm understand the structure and meaning of sentences.

- *Named Entity Recognition (NER):* NER is the process of identifying and categorizing entities (e.g., people, organizations, locations) mentioned in the text, allowing the virtual assistant to understand the context and relationships between entities.

- *Sentiment Analysis:* Sentiment analysis involves determining the emotional tone or sentiment expressed in the text, enabling the virtual assistant to provide empathetic and personalized responses.

## 2.3 Speech recognition

Speech recognition is the technology that enables computers to convert spoken language into text. Advanced virtual assistants use speech recognition to understand and process voice commands, allowing users to interact with them using natural speech.

There are two main approaches to speech recognition:

- *Acoustic Modeling:* Acoustic modeling involves creating a mathematical representation of the relationship between the acoustic features of speech and the corresponding linguistic units (e.g., phonemes). This allows the system to predict the most likely sequence of linguistic units given the acoustic features of the speech.

- *Language Modeling:* Language modeling involves estimating the probability of a sequence of words or linguistic units occurring in each context. This information is combined with the acoustic model to generate the most likely speech transcription.

### 2.4 Knowledge representation and reasoning

Advanced virtual assistants must represent, store, and manipulate knowledge to answer questions and perform tasks effectively. Knowledge representation and reasoning techniques allow virtual assistants to organize information, draw inferences, and make decisions based on available knowledge.

Some common knowledge representation and reasoning techniques include:

- *Ontologies:* Ontologies are formal representations of knowledge that define a domain's concepts, relationships, and properties. They provide a structured way for virtual assistants to represent and reason about the world.

- *Semantic Networks:* Semantic networks are graph-based representations of knowledge, where nodes represent concepts or entities, and edges define relationships between them. Semantic networks allow virtual assistants to represent complex relationships and reason about the connections between different pieces of information.

- *Rule-Based Systems:* Rule-based systems use a set of predefined rules or heuristics to represent knowledge and perform reasoning. Virtual assistants can use these rules to infer new information or make decisions based on the available data.

### 2.5 Integration of AI technologies

Advanced virtual assistants rely on integrating multiple AI technologies and techniques to provide a seamless and intelligent user experience. This includes the combination of machine learning, NLP, speech recognition, and knowledge representation and reasoning to understand, process, and respond to user requests effectively.

Integration of these technologies enables advanced virtual assistants to:

- Understand the user's intent and context, allowing them to provide more accurate and relevant information.

- Generate human-like responses and carry out complex tasks using natural language commands.

- Learn from user interactions and adapt their behavior, resulting in a more personalized and intuitive user experience.

## 3. Challenges and opportunities in advanced virtual assistant development

Despite advanced virtual assistants' tremendous progress and potential, several challenges must be addressed to unlock their full potential. We will explore some of the critical challenges developers and researchers face in advanced virtual assistants and discuss the opportunities that arise from overcoming these challenges [3].

### 3.1 Context awareness and understanding

One of the significant challenges in developing advanced virtual assistants is enabling them to understand and respond to user requests in a context-aware manner. This requires recognizing and considering the context in which a conversation occurs, including the user's intent, preferences, and history of interactions.

*Opportunity:* Advanced virtual assistants can provide more accurate and relevant information by improving context awareness and understanding, resulting in a more personalized and intuitive user experience.

### 3.2 Continuous learning and adaptation

Advanced virtual assistants must continually learn and adapt to new information and changing user needs. However, current machine learning techniques often require large amounts of labeled data and significant computational resources, making implementing continuous learning and adaptation in real time challenging.

*Opportunity:* Developing new machine learning techniques and algorithms that enable continuous learning and adaptation will allow advanced virtual assistants to evolve and improve over time, providing users with increasingly personalized and intelligent assistance.

### 3.3 Multimodal interaction

While most virtual assistants currently rely on text or voice-based interactions, there is a growing need for multimodal interaction that combines various input and output modalities, such as text, speech, gestures, and facial expressions.

*Opportunity:* By incorporating multimodal interaction, advanced virtual assistants can provide a more natural and engaging user experience, enabling users to interact with technology more intuitively and seamlessly.

### 3.4 Scalability and resource efficiency

As advanced virtual assistants become more complex and capable, they also require more computational resources to process and analyze data. This can lead to scalability challenges and increased energy consumption, making it difficult to deploy these systems on a large scale.

*Opportunity:* Addressing the scalability and resource efficiency challenges will enable the widespread adoption of advanced virtual assistants, making these intelligent systems more accessible and affordable for users worldwide.

### 3.5 Interoperability and standardization

With the growing number of virtual assistant platforms and devices, there is a need for greater interoperability and standardization to ensure seamless integration and compatibility between different systems.

*Opportunity:* Establishing industry standards and promoting interoperability will facilitate the development of more cohesive and integrated virtual assistant ecosystems, enabling users to access and manage multiple services and applications through a single, unified platform.

## 4. Envisioning the future of advanced virtual assistants

Advanced virtual assistants have the potential to revolutionize the way we interact with technology and reshape various aspects of our lives. We will envision the future of advanced virtual assistants, discussing cutting-edge advancements and emerging trends that could redefine human-computer interaction and transform the world as we know it [4].

### 4.1 Seamless integration and ubiquity

In the future, advanced virtual assistants will become even more seamlessly integrated into our daily lives, with a growing number of connected devices, platforms, and environments incorporating intelligent assistance capabilities. From wearable devices and smart appliances to public spaces and transportation systems, virtual assistants could become ubiquitous, providing personalized, context-aware assistance whenever and wherever needed [5].

### 4.2 Advanced multimodal interaction

Future virtual assistants are expected to incorporate more advanced multimodal interaction capabilities, allowing users to communicate with them using a variety of input and output modalities, such as text, speech, gestures, and facial expressions. This will enable more natural, intuitive, and immersive interactions, bridging the gap between humans and machines [6].

### 4.3 Enhanced cognitive capabilities

As AI research and development advances, we can expect future virtual assistants to exhibit enhanced cognitive capabilities, including improved reasoning, problem-solving, and creativity. These advanced cognitive skills will enable virtual assistants to understand complex human needs better and provide more sophisticated, insightful, and proactive assistance.

### 4.4 Collaborative intelligence

In the future, advanced virtual assistants may move beyond individual assistance to facilitate collaborative intelligence, working with other virtual assistants and human users to solve problems, make decisions, and generate new ideas. This collaborative approach could help unlock new collective intelligence, innovation, and productivity levels.

### 4.5 Ethical and responsible AI

As advanced virtual assistants evolve and integrate into our lives, ethical and responsible AI development will become increasingly paramount. Future virtual assistants must be designed with ethical considerations, incorporating privacy-preserving technologies, empathy and emotional intelligence, and equitable access for all users.

## 5. Conclusion

In conclusion, Advanced Virtual Assistants have revolutionized how we interact with technology and simplify our lives in many ways. From simple voice-activated commands to more complex tasks like scheduling appointments and managing our finances, these intelligent assistants have become essential to our daily routine. As technology evolves, we can expect to see even more advanced virtual assistants that can seamlessly integrate with our smart homes and other devices, making our lives even more convenient. However, it is essential to recognize that these assistants also raise significant ethical and privacy concerns that must be addressed by both users and developers alike. Overall, the rise of Advanced Virtual Assistants marks an exciting new chapter in the ongoing evolution of technology, and we cannot wait to see what the future holds.

## Author details

Ali Soofastaei
Innovative AI (IAI), Brisbane, Australia

*Address all correspondence to: ali@soofastaei.net

## References

[1] Baek S et al. Implementation of a virtual assistant system based on deep multi-modal data integration. Journal of Signal Processing Systems (South Korea). 2023;**2023**:1-11

[2] Gnana Rajesh D, Tamilarasi G, Khan ME. Voice and text-based virtual assistant for academic advising using knowledge-based intelligent decision support expert system. In: Sentiment Analysis and Deep Learning: Proceedings of ICSADL 2022. Australia: Springer; 2023. pp. 483-491

[3] Mala DJ. Machine learning-based intelligent assistant for smart healthcare. In: Machine Learning and Artificial Intelligence in Healthcare Systems. UK: CRC Press; 2023. pp. 265-286

[4] Ponnusamy SS et al. IBIS: An interactive virtual assistant for system engineers. In: AIAA SCITECH 2023 Forum. USA; 2023

[5] Qian C et al. Distinguish sense from nonsense: Out-of-scope detection for virtual assistants. arXiv Preprint arXiv:2301.06544. 2023

[6] Ramdani CMS, Rachman AN, Dewi ENF. Application of virtual assistant in information system for student practicum case study Laboratory Informatics Department Siliwangi University. CESS (Journal of Computer Engineering, System and Science) (USA). 2023;**8**(1):47-59

Chapter 2

# ICT and Physical Activity

*Hideyuki Namba*


## Abstract

Recently, physical inactivity has emerged as a problem worldwide and the effective application of digital health using information and communication technology (ICT) has been focused on as a potential solution to this problem. An overview of research using web-based physical activity assessment systems and wearable devices is presented. The following three topics will be discussed: (1) mobile health market and sensing technology, (2) evaluation of physical activity using ICT, and (3) a study of physical activity promotion through digital intervention. Wearable devices (a generic term for information terminals that are worn and carried around) have made remarkable progress in recent years, and physical activity promotion and weight loss interventions using PCs (personal computers) and mobile devices are considered to have advantages in terms of human and economic costs compared to traditional face-to-face interventions. A wide variety of data will be collected using various wearable devices, and artificial intelligence (AI) technologies such as machine learning and deep learning will be incorporated to develop applications that introduce future risk projections and other information. Rather than people catching up with technological advances, it is important to take a viewpoint of how technology can be used to enrich people's lives.


**Keywords:** physical activity, smartphones, voice recognition, digital intervention, m-Health

## 1. Introduction

In recent years, lifestyle changes resulting from technological developments in the work and transportation environment have led to widespread physical inactivity in people worldwide [1]. As long sedentary behavior and physical inactivity are respectively associated independently with risk factors for noncommunicable diseases such as diabetes, coronary heart disease, and some cancers, in 2020, WHO issued guidelines for physical activity and sedentary behavior, recommending at least 150–300 minutes of moderate-intensity aerobic physical activity throughout the week for adults, at least 75–150 minutes of high-intensity aerobic physical activity, and at least 2 minutes of muscle-strengthening exercise and less sedentary behavior per week for adults [2].

Although the ecological approach [3] such as community intervention, sports promotion, health education, and environmental improvements such as maintenance of the parks and walkways are needed to promote physical activity among a large number of people, efforts to promote physical activity using ICT represented by smartphones, which have been remarkably developed and spread in recent years [4], have been the focus of public interest. However, there are a variety of physical activity intervention methods using online services, and the effectiveness of these methods is not clearly described. In other words, clinical research has not caught up with technological advances. This paper reviews the research to the present and summarizes the future needs of the society.

## 2. Mobile health market and sensing technology

Healthcare management, medical treatment, and medical support using smartphones and tablets for health-related information are known collectively as mobile health (m-Health). The global m-health market is growing rapidly, from $4.5 billion in 2013 to $23 billion in 2017 [5] and from $36.43 billion in 2021 to $113.25 billion in 2026 [6]. The use of the Internet and mobile devices, especially SMS, has been shown to be a cost-effective intervention to promote physical activity for a large population [7]. Systematic reviews have reported a positive effect mainly in Europe and North America [8, 9].

Recently, wearable devices (a general term for information terminals that are carried around with the wearer) have made remarkable progress, and some representative examples include wristwatch-type smart watches, glasses-type smart glasses, smart rings with sensors mounted on them, smart contact lenses, and even skin-attached sensors for pulse monitoring of pulse rate by skin-attached sensors [10] has also been reported as a wearable device. The main sensing methods used in wearable devices include acceleration, gyro, light, infrared, bioelectrical potential, and GPS. The three main categories of applications are (1) information on the body and mind (blood flow, heart rate, body temperature, EEG, eye movement), (2) information on position and speed (location information, movement information), and (3) input (operation) and motion support (body movement, muscle movement) [11].

Health-related data can be collected using sensors installed in the smartphone itself. The wearable devices mentioned previously can be used to collect a wide variety of data from multiple sources, capture it on a smartphone *via* Bluetooth, and manage the data using applications. By utilizing a web-based application, data can be stored and processed on a cloud server *via* a 4G (LTE) network. By using the cloud, service providers have the advantage of expanding the fields in which they can create new and various services, and users can check and manage visualized data such as physical activity, sleep status, stress checks, and dietary calories.

Active lifestyle advice based on observational data, including elements of behavioral science and gamification, feedback content to increase user adherence, and social networking service (SNS)-based community sites could work on human cognition, arouse motivation, trigger physical activity, and contribute to the prevention of lifestyle-related diseases caused by physical inactivity.

## 3. Evaluation of physical activity using ICT

Up to the present, various methods have been developed to assess physical activity, such as questionnaires, accelerometers, and more recently Global Positioning

System (GPS), depending on the purpose of use. However, there is little evidence for the promotion of physical activity for a large population of several tens of thousands of residents. This is because accelerometers are difficult to widely distribute to a large number of people due to their cost, and self-reports have low validity and are not suitable for longitudinal evaluation of the effects of interventions. Conventional physical activity measurement methods based on questionnaires have the advantage of measuring physical activity in a large number of subjects, but they have limitations in measurement accuracy. The correlation coefficient is about 0.3–0.6 when compared to Doubly-Labeled Water (DLW), which is the gold standard method for measuring total energy expenditure (TEE) [12].

The validity of self-reported physical activity measurement methods using the web and cell phones has been reported in several studies. A web-based questionnaire called Active-Q was developed to select the intensity of each activity from four categories (work, transportation, leisure, and sports), and its validity was verified using the DLW method on 37 general adult subjects, and a significant association of $r = 0.52$ ($p < 0.001$) was reported for TEE [13]. Thirty men and women with cardiovascular diseases were asked to self-report their activity levels during the day and at night using a smartphone-based system and validated with an accelerometer, and an association of $r = 0.45$ ($p < 0.05$) was reported in physical activity levels [14]. These self-reported physical activity measurements using the web or cell phones are superior in that they can simultaneously assess the physical activity of many people at the same time, but their measurement accuracy is similar to that of traditional questionnaires, and the content of the questions is not well designed. On the other hand, a study has been published in which a high correlation of 0.88 was found between TEE using the DLW method and the physical activity measurement method using the paper-based 24-hour recall method [15]. The validity of this measurement method for assessing TEE exceeds the accuracy of accelerometers when the DLW method is used as the standard. The paper-based method is difficult to use for a large number of people due to the lack of interactive responses and difficulty in memory recall.

A physical activity system using IT devices based on the 24-hour look-back method has been developed, and its validity has been shown to correlate with TEE by the DLW method with $r = 0.874$ ($p < 0.001$) and with activity energy expenditure (AEE) with $r = 0.679$ ($p < 0.001$) [16]. It is reported that the measurement accuracy is equivalent to that of a 3-axis accelerometer in estimating TEE. A feature of this system is the method of dragging and dropping illustrations on the web screen to fill in the timeline (**Figure 1**). Using such a system, it is possible to assess the physical activity of a large number of people simultaneously and at the same time with high accuracy at low cost [17] and to assess each activity intensity (**Figure 2**) [18]. It is also possible to evaluate the physical activity of a large population with COVID-19 online, and the effects of COVID-19 on physical activity have been clarified [19, 20]. The 24-hour activity recording method suggests the possibility of physical activity assessment by voice input (**Figures 3** and **4**) by introducing a voice recognition application programming interface (API) [21].

## 4. A study of physical activity promotion through digital intervention

Literature studies on e-health and m-health regarding physical activity, sedentary behavior, and diet [22] reported an average annual increase of 26% in the number of articles between 2000 and 2016, with a rapid increase since 2014. A review [23] that

**Figure 1.**
*Screenshot of the web-based simplified physical activity record system [17].*

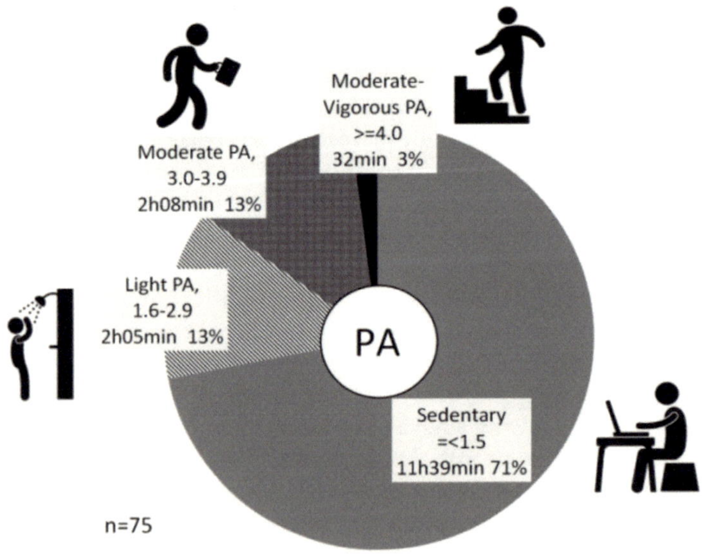


**Figure 2.**
*Activity time per activity intensity [18].*

conditioned digital interventions on healthy lifestyles cited 107 articles from 2015 to 2020 and found that methods for successful digital interventions include monitoring, motivation, goal setting, personalized feedback, participant engagement, psychological empowerment, persuasion, digital literacy, self-efficacy, and authenticity.

Forty review articles published up to March 2021 on digital interventions to promote physical activity were analyzed [4]. Physical activity was assessed by objective methods (wearables and smartphone active trackers) in 30%, objective methods

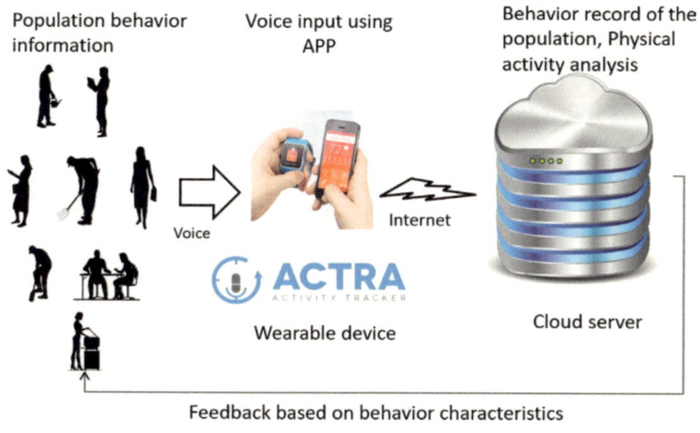


**Figure 3.**
*A behavior-recording system with a voice input app [21].*

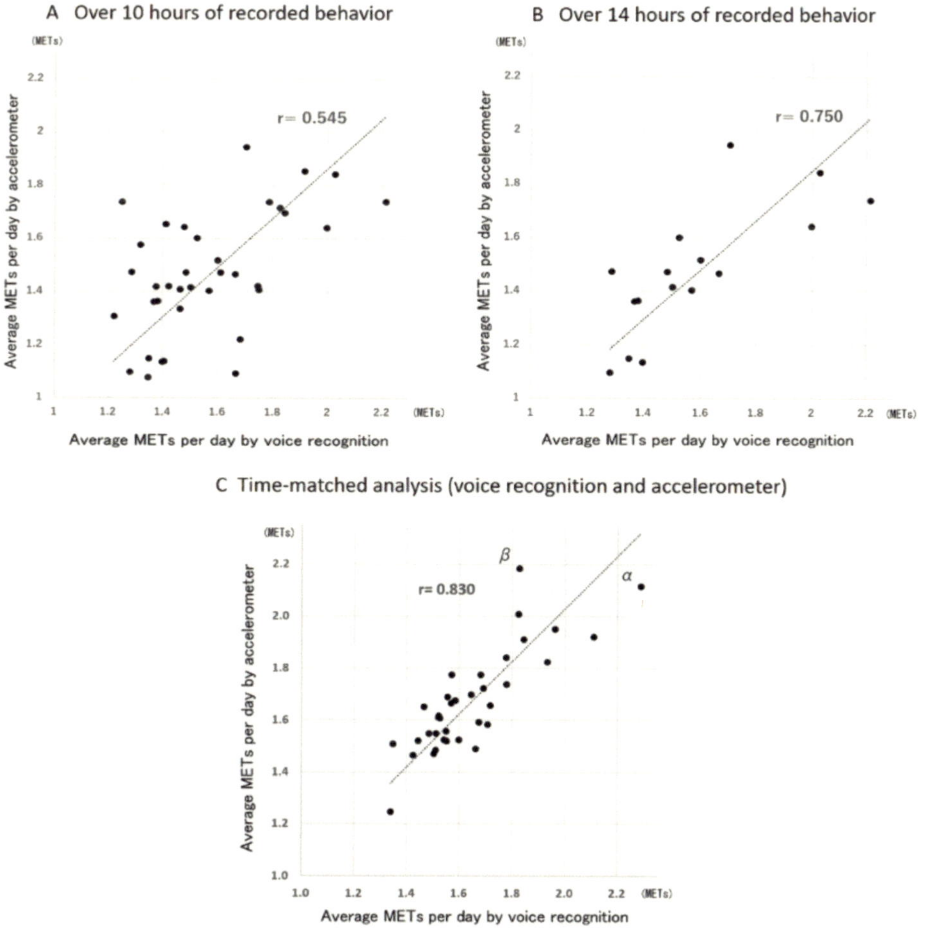


**Figure 4.**
*Correlation of average measured METs with voice recognition app [21].*

plus self-report in 60%, and evaluation based on behavioral change theory, including goal setting, self-monitoring, and feedback as a theoretical framework, in 55% of the cases. In order to clarify the effectiveness of digital interventions for physical activity promotion, the need to introduce an evaluation framework at the system development stage is noted.

## 5. Introducing advanced research using smart shoe sensors

Smart shoes are Internet of Things (IoT) shoes with built-in motion sensors and other sensors in the shoes and AI. These sensors, which include accelerometers, gyros, and GPS, can evaluate walking and running paces, landing times, pronation, stride, pitch, and landing impact force during movement. It is important to evaluate the quality of walking because walking is a main part of physical activity for people living a typical life. Good gait quality means that a person requires less energy to move the same distance. In other words, if the mechanical efficiency of walking (= total work/energy expenditure) can be evaluated using smart shoes, the quality of each individual's gait could be assessed and improved [24].

A walking test was conducted on 35 middle-aged and elderly subjects to clarify the relationship between mechanical energy efficiency in walking and the level of physical activity. The subjects performed three types of walking, normal walking, fast walking, and slow walking, on an approximately 10-m walking path for five strides each, and the energy efficiency of one walking cycle was evaluated using a force plate. At the same time, smart shoes (with a built-in small motion sensor) and plantar pressure distribution were evaluated and assessed to clarify whether the energy efficiency of walking could be evaluated even without a force plate (**Figure 5**). Furthermore, we evaluated energy expenditure by DLW for 1 week and physical activity by the triaxial accelerometer, in order to clarify the relationship between these data and the mechanical efficiency of walking (= total workload/energy expenditure).

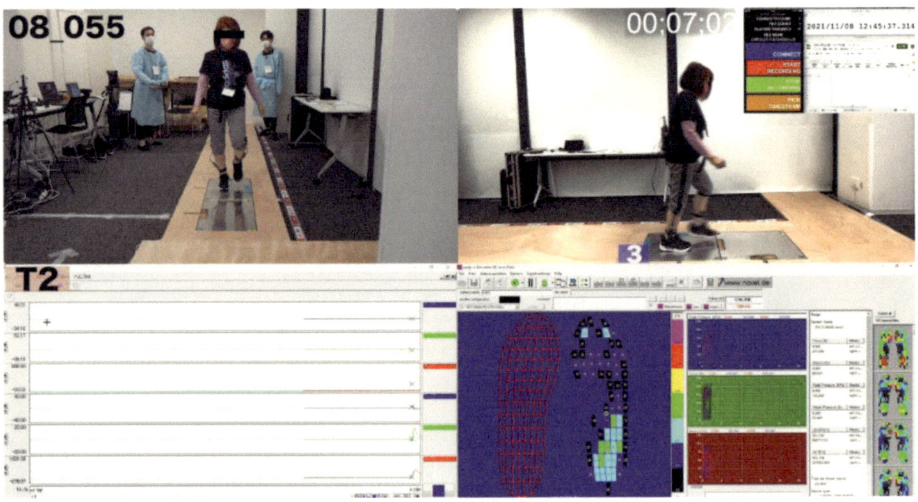


**Figure 5.**
*Basic data collection experiment using smart shoes.*

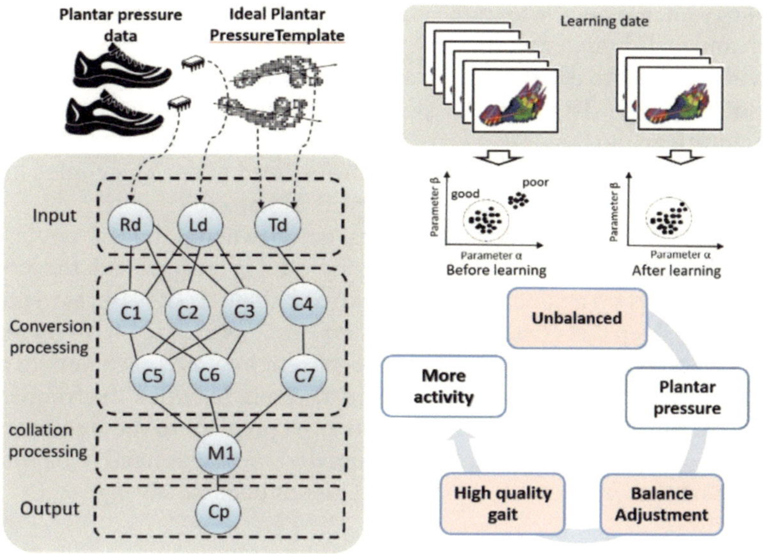


**Figure 6.**
*AI-based feedback system using smart shoes.*

A schema of effective algorithms for promoting physical activity *via* AI is shown for data obtained from smart shoes (**Figure 6**). Deep learning is incorporated using an application programming interface (API) in order to link with the software that has been developed so far. Data obtained from the smart shoes are transformed through the conversion process (C1–C7) and then matched (M1) to generate a program that determines the validity of the data using deep learning technology. Accuracy management of wearable sensors is the key to evaluating walking behavior with wearable sensors (smart shoes) and generating tailor-made advice provided by AI. The key point is to evaluate mechanical energy with high accuracy. Recent measurement technology has advanced dramatically, and there is a possibility of evaluating the energy efficiency of walking with a certain level of accuracy using smart shoes, which are linked to a smartphone and sensors and are considered to be capable of inexpensive, high-precision motion analysis.

The final goal is to develop an AI feedback system using smart shoes based on data obtained during walking. If it is possible to increase physical activity within the optimal range for each individual, it could lead to the prevention of lifestyle-related diseases such as diabetes and hypertension, and contribute to solving the various problems confronting an aging society.

## 6. Conclusions

The history of digital intervention research using online systems is short, and technology is constantly advancing, but it is not always up to date with people's needs. JMIR Publications, a research journal dedicated to digital health, launched in 1999, has a vision: "We believe that people should be able to make effective, well-informed decisions through health-related research and technology, and that they should be able to make the best use of technology. We envision a world where health research and technology enable people to make effective, informed decisions, self-manage their health and well-being, and live happier, healthier lifestyles [25]. In the future, this field will continue to develop applications that use various wearable devices to collect a wide variety of data and incorporate artificial intelligence (AI) technologies such as machine learning and deep learning to introduce predictions of future risks. It is important to consider how technology can be utilized to enrich people's lives, rather than how people can catch up with technological advances.

Finally, as proposals for increasing physical activity using the ICT environment, the following four approaches should be considered: (1) an approach to elements that can be changed based on an analysis of individual behavior, (2) efforts to prevent health problems (sleep disorders, stiff shoulders, etc.) associated with increased use of PCs, smartphones, and other devices, (3) a system in which each person can choose the frequency of ICT-based interventions, and a system in which the frequency of ICT-based interventions can be adjusted as much as possible to the frequency of face-to-face meetings with ICT users, and (4) automated messages need to be moderated to the extent that they are not unpleasant and do not become boring.

## Acknowledgements

I would like to thank my collaborators for their active support of the study and the subjects who cooperated in the research. This work was supported by JSPS KAKENHI 17 K01800 and 21 K11478.

## Conflict of interest

The authors declare no conflict of interest.

## Author details

Hideyuki Namba
Physical Education Laboratory, College of Science and Technology Nihon University, Chiba, Japan

*Address all correspondence to: nanba.hideyuki@nihon-u.ac.jp

## References

[1] World Health Organization. Global Action Plan on Physical Activity 2018-2030: More Active People for a Healthier world. Geneva: World health Organization; 2018

[2] WHO Steering Group. WHO Guidelines on Physical Activity and Sedentary Behavior. Geneva: World Health Organization; 2020

[3] Sallis JF, Cervero RB, Ascher W, Henderson KA, Kraft MK, Kerr J. An ecological approach to creating active living communities. Annual Review of Public Health. 2006;**27**:297-322

[4] De Santis KK, Jahnel T, Matthias K, Mergenthal L, Al Khayyal H, Zeeb H. Evaluation of digital interventions for physical activity promotion: Scoping review. JMIR Public Health and Surveillance. 2022;**8**(5):e37820. DOI: 10.2196/37820

[5] Touching Lives through Mobile Health Assessment of the Global Market Opportunity. PricewaterhouseCoopers; 2012

[6] mHealth Apps Global Market Report 2022. The Business Research Company. No.5 and No.6 are India and U.S.A. receptivity; 2022

[7] Pratt M, Sarmiento OL, Montes F, Ogilvie D, Marcus BH, Perez LG, et al. The implications of megatrends in information and communication technology and transportation for changes in global physical activity. Lancet. 2012;**380**:282-293. DOI: 10.1016/S0140-6736(12)60736-3

[8] Webb TL, Joseph J, Yardley L, Michie S. Using the internet to promote health behavior change: A systematic review and meta-analysis of the impact of theoretical basis, use of behavior change techniques, and mode of delivery on efficacy. Journal of Medical Internet Research. 2010;**12**(1):e4. DOI: 10.2196/jmir.1376

[9] Stephens J, Allen J. Mobile phone interventions to increase physical activity and reduce weight: A systematic review. The Journal of Cardiovascular Nursing. 2013;**28**(4):320-329. DOI: 10.1097/JCN.0b013e318250a3e7

[10] Kim Y, Suh JM, Shin J, Liu Y, Yeon H, Qiao K, et al. Chip-less wireless electronic skins by remote epitaxial freestanding compound semiconductors. Science. 2022;**377**:859-864. DOI: 10.1126/science.abn7325

[11] Ministry of Internal Affairs and Communications. Teaching Materials for Comprehensive Acquisition of ICT Skills. Data Collection Technology and Wearable Devices. Ministry of Internal Affairs and Communications in the Government of Japan. Available from: https://norikoe.net/wp-content/uploads/2020/01/ict_skill_all_set.pdf [Accessed: May 22, 2023]

[12] Neilson HK, Robson PJ, Friedenreich CM, Csizmadi I. Estimating activity energy expenditure: How valid are physical activity questionnaires? The American Journal of Clinical Nutrition. 2008;**87**(2):279-291. DOI: 10.1093/ajcn/87.2.279

[13] Bonn SE, Trolle Lagerros Y, Christensen SE, Möller E, Wright A, Sjölander A, et al. Active-Q: Validation of the web-based physical activity questionnaire using doubly labeled water. Journal of Medical Internet Research. 2012;**14**(1):e29. DOI: 10.2196/jmir.1974

[14] Pfaeffli L, Maddison R, Jiang Y, Dalleck L, Löf M. Measuring physical activity in a cardiac rehabilitation population using a smartphone-based questionnaire. Journal of Medical Internet Research. 2013;**15**(3):e61. DOI: 10.2196/jmir.2419

[15] Koebnick C, Wagner K, Thielecke F, Moeseneder J, Hoehne A, Franke A, et. al. Validation of a simplified physical activity record by doubly labeled water technique. International Journal of Obesity. 2005;**29**(3):302-309. DOI: 10.1038/sj.ijo.0802882

[16] Namba H, Yamaguchi Y, Yamada Y, Tokushima S, Hatamoto Y, Sagayama H, et al. Validation of web-based physical activity measurement systems using doubly labeled water. Journal of Medical Internet Research. 2012;**14**(5):e123. DOI: 10.2196/jmir.2253

[17] Namba H, Yamada Y, Ishida M, Takase H, Kimura M. Use of a web-based physical activity record system to analyze behavior in a large population: Cross-sectional study. Journal of Medical Internet Research. 2015;**17**(3):e74. DOI: 10.2196/jmir.3923

[18] Namba H, Kurosaka Y, Minato K, Yamada Y, Kimura M. Validation of a web-based physical activity measurement system using a tri-axial accelerometer. Research in Exercise Epidemiology. 2015;**17**(1):19-28. DOI: 10.24804/ree.17.19

[19] Yamada Y, Namba H, Date H, Kitayama S, Nakayama Y, Kimura M, et.al. Regional difference in the impact of COVID-19 pandemic on domain-specific physical activity, sedentary behavior, sleeping time, and step count: Web-based cross-sectional Nationwide survey and accelerometer-based observational study. JMIR Public Health and Surveillance 2023;**9**:e39992. DOI: 10.2196/39992

[20] Namba H, Kita T, Kobayashi K, Kimura M. Measurement of physical activity using the WEB for university students under the spread of COVID-19 infection. Japanese Journal of Physical Education and Sport for Higher Education. 2023;**20**:23-32

[21] Namba H. Physical activity evaluation using a voice recognition app: Development and validation study. JMIR Biomed Eng. 2021;**6**(1):e19088. DOI: 10.2196/19088

[22] Müller AM, Maher CA, Vandelanotte C, Hingle M, Middelweerd A, Lopez ML, et al. Physical activity, sedentary behavior, and diet-related eHealth and mHealth research: Bibliometric analysis. Journal of Medical Internet Research. 2018;**20**(4):e122. DOI: 10.2196/jmir.8954

[23] Chatterjee A, Prinz A, Gerdes M, Martinez S. Digital interventions on healthy lifestyle management: Systematic review. Journal of Medical Internet Research. 2021;**23**(11):e26931. DOI: 10.2196/26931

[24] Namba H. Development of Walking Motion Learning Support System by Artificial Intelligence using Wearable Sensors. National Institute of Infomatics in JAPAN. Available from: https://kaken.nii.ac.jp/en/grant/KAKENHI-PROJECT-21K11478/ [Accessed: May 23, 2023]

[25] Eysenbach G. Celebrating 20 years of open access and innovation at JMIR publications. Journal of Medical Internet Research. 2019;**21**(12):e17578. DOI: 10.2196/17578

Chapter 3

# Human-AI Collaboration for Smart Education: Reframing Applied Learning to Support Metacognition

*James Hutson and Daniel Plate*

## Abstract

This chapter investigates the profound influence of intelligent virtual assistants (IVAs) on the educational domain, specifically in the realm of individualized learning and the instruction of writing abilities and content creation. IVAs, incorporating generative AI technologies such as ChatGPT and Stable Diffusion, hold the potential to bring about a paradigm shift in educational programs, emphasizing the enhancement of advanced metacognitive capacities rather than the fundamentals of communication. The subsequent recommendations stress the need to cultivate enduring proficiencies and ascertain tailored learning approaches for each learner, which will be indispensable for success in the evolving job market. In this context, prompt engineering is emerging as a vital competency, while continuous reskilling and lifelong learning become professional requisites. The proposed innovative method for teaching writing skills and content generation advocates for a reconfiguration of curricula to concentrate on applied learning techniques that accentuate the value of contextual judgment as a central pedagogical tenet and the mastery of sophisticated metacognitive abilities, which will be pivotal in the future of work.

**Keywords:** metacognition, Chatbot, ChatGPT-4, intelligence virtual assistants (IVA), intelligent personal assistant (IPA)

## 1. Introduction

Intelligent virtual assistants (IVAs) or intelligent personal assistants (IPAs) represent software agents capable of executing tasks or providing services for individuals based on directives or inquiries [1]. These encompass more recent advances, including generative artificial intelligence (AI) technologies such as ChatGPT, Stable Diffusion, Notion, and others. The enhanced capabilities of these AI tools have profoundly disrupted the educational landscape, surpassing any other emergent technology of the new millennium in speed and scale of adoption [2]. At the same time, some have countered that investors fundamentally misunderstand the technology behind language models like ChatGPT and OpenAI-powered Bing Search. For instance, Harrison [3] notes that although these AI models sound impressively human, they do not synthesize information, nor do they provide thoughtful, analytical, or accurate answers. Rather, they function similarly to predictive text features, merely predicting the next words in a sentence based on probability.

Nevertheless, these generative technology tools have already made inroads in industry and education. For instance, surveys of one-thousand business leaders less than a month after the launch of the stable release of ChatGPT-3 in early February, found that nearly 50% of their companies have implemented the chatbot, with half of these companies reporting that the chatbot has replaced some workers. Williams [4] continues to report that companies use such automated assistants for various tasks, including writing code (66%), copywriting and content creation (58%), customer support (57%), and meeting summaries (52%). In the hiring process, the chatbot is utilized for writing job descriptions (77%), drafting interview requisitions (66%), and responding to applications (65%). With so many applications, there is little wonder why this new technology has made its way beyond the office and into the classroom, as well.

Given the fact that AI continually learns from accumulated historical data and user interactions, the potential for customized learning pathways is demonstrable, promising to revolutionize pedagogical methods across all levels of education. Furthermore, the generative capacity of these digital assistants has the capacity to grant increased access to education for underrepresented populations and individuals with learning disabilities, promoting a more equitable educational landscape [5]. At the same time, learning how to interact with AI tools will become a key job skill moving forward. Prompt engineering and generation has already emerged as one of the top 10 skills necessary for maintaining a competitive edge in the industry [6]. Therefore, these new educational avatars not only offer a revolutionary new approach to pedagogy, but their very use will be central to students' future personal and professional lives. Preparing them with these new tools should be a priority for all educators.

In order to address the pressing need, this chapter will investigate the potential applications of IVAs, particularly generative AI as educational learning companions and tutors. Practical strategies for integration into current and future educational programs shall be addressed, and a reconceptualization and recentering of metacognition (learning how to learn), as the most valuable transferable and durable skill students can now gain from education presented. Given the advent of what Clark [7] has coined 'PedAIgogy,' technology is no longer utilized for accessing knowledge but also for creating, modifying, organizing, synthesizing, and evaluating it. The new pedagogical approach involves co-creating multimedia content and engaging in complex relationships with via collaborative virtual agents. With these new automatable abilities, there will be a return to the Socratic approach to knowledge acquisition and application that will foster the development, analysis, and evaluation of co-created literature, imagery, audio, and video, as well as new research, art, teaching, and learning methods [8].

More specifically, one may take as an example how the teaching and learning of the writing process will evolve given these new capabilities. Given that a primary objective in composition coursework has been grounded in instruction of grammar, syntax, and sentence-level writing, these courses tend to emphasize the importance of clear, concise, and grammatically correct writing at the sentence level [9]. Although the standard freshman composition sequence often assumes basic proficiency at the sentence-level, the practical reality is that the grading of composition papers often places too much stress on grammar/usage errors, and in-course remedies for gaps in grammar knowledge take up at least some valuable instruction time. If students do gain a good grasp of the basics, they are sometimes able to focus on more advanced sentence-level writing skills like sentence style and rhetorical devices. Practicing writing clear, concise, and grammatically correct sentences through various assignments is often used to reinforce these skills. Feedback is the last step in the iterative process either via peers or instructors [10].

However, given the new abilities of chatbots and digital assistants to generate drafts on any topic with the right prompting, the pedagogical focus should shift towards nurturing students as editors, rather than sentence generators. The paradigm shift has been clearly seen over the past few years where individuals or even trained professionals are increasingly ceasing to be content creators. With the wide availability of video and audio editing software on smartphones, the ease of content creation and distribution has already disrupted the media and content industry. The growth of mobile video, user-generated content, and social networks combined with advancements in AI-powered content creation, dissemination, and interaction technologies already allows for the production of high-quality content in various forms, including video, music, and augmented reality (AR), which can be rapidly shared with a vast global audience [11]. But now AI-powered tools have evolved and also now have the ability to create sophisticated written content. Therefore, these new digital assistants will assume the responsibility of generating sentence-level content, thereby producing initial drafts that humans will subsequently refine through editing. Education, particularly at the postsecondary level, must reorient the training of students to embrace their new roles as editors in this new human-AI collaborative relationship.

The integration of generative AI into education has the potential to revolutionize the way students learn and teachers teach, with the use of personalized services and individualized learning modalities becoming increasingly mainstream. However, this widespread adoption of AI in education requires adjustments to classroom practices, assessment methods, and credentialing. The importance of critical thinking, digital and information fluency, and fact-checking is becoming more crucial, and new jobs such as prompt engineering have emerged. Metacognition plays a crucial role in education, and educators can utilize various strategies to foster critical thinking and metacognitive skills. With the integration of generative AI tools like ChatGPT, teachers can create personalized and engaging learning experiences that encourage metacognitive reflection and standardize the development of critical thinking skills. The use of AI in teaching can contribute to a more effective and reflective learning environment, leading to better outcomes for students.

This chapter focuses on the paradigm shift required in writing instruction, where editing and prompt development become vehicles for enhancing metacognition, and a novel model for writing education can be envisioned where students receive tutoring while editing, ultimately leading to improved metacognitive thinking, sentence-level writing, and tailored feedback and support. The following sections will outline the future of education, importance of the development of metacognitive skills and new strategies for learning how to write.

- **Future of Education** outlines how the integration of generative AI into education is poised to revolutionize the way students learn and teachers teach. Institutions of higher education are adapting to changing societal and economic demands, while the use of AI-driven personalized services, and individualized learning modalities are becoming more mainstream. At the same time, the widespread adoption of AI in education requires adjustments to classroom practices, assessment methods, and credentialing. Finally, new jobs such as prompt engineering have appeared, highlighting the potential for innovation and growth in the field of education.

- **Metacognition in Education** focuses on how metacognition plays a crucial role in education, enabling learners to become self-aware, reflective, and effective thinkers. Examples of activities are discussed, such as constrained choice exercises and metacognitive prompts, that educators can utilize to foster critical thinking and metacognitive skills. Moreover, the integration of generative AI tools like ChatGPT can create personalized and engaging learning experiences that encourage metacognitive reflection and standardize the development of critical thinking skills. Through such tools, educators can provide individualized feedback, generate tailored learning resources, facilitate group discussions, monitor student progress, and evaluate their own teaching practices by incorporating them into the classroom.

- **Writing Education and AI** outlines the integration of AI into education and how that requires a paradigm shift in writing instruction that emphasizes editing and prompt development as vehicles for enhancing metacognition. Students must develop language skills that enable effective interaction with AI and facilitate learning from AI-generated content in order to ensure a competitive advantage in the future of work. In order to support this process, a novel model for writing education where students receive tutoring while editing will be introduced. In this approach, AI generates preliminary material for a paper, and students engage in dialog with the AI to understand the rationale behind specific choices. Through utilizing AI as a tutor and editor, metacognitive thinking and improve sentence-level writing can be developed, while also providing students with tailored feedback and support.

## 2. Future of education

In response to evolving societal and economic demands, higher education is witnessing significant trends aimed at enhancing student outcomes, employability of graduates, and institutional sustainability. Hanover Research, for instance, identifies five critical trends shaping this landscape [12]. Among these trends is the modernization of academic programs with the growing demand to incorporate career-oriented, stackable structures that closely align with the demands of the job market, thus fostering the development of employable skills among graduates [13–14]. Simultaneously, there is a growing demand for debt-free education, prompting colleges and universities to scrutinize alternative funding models to mitigate the financial burden on students [15]. On the other hand, the higher education sector is increasingly prioritizing inclusivity and support services, aiming to cultivate a welcoming environment that promotes acceptance, belonging, and achievement for all students [16–17]. These services often increase tuition fees when filled by human agents, but enhance retention rates and optimize overall outcomes, ultimately contributing to the provision of accessible and personalized education. Finally, as competition in the market intensifies, institutions are under growing pressure to deliver tangible results, leading to the adoption of data-driven approaches for resource allocation and decision-making, emphasizing the need for measurable outcomes in the educational sphere [18].

These disruptive trends affecting institutions of higher education seem to be placing demands on decision-makers that are at odds with one another. On the one hand, greater individualized support is demanded with greater accessibility and individualized learning that lead to measurable outcomes and job placement for graduates.

On the other hand, more affordable paths to a college degree make it necessary to look for solutions beyond what has been attempted in the past, which would be to hire more faculty and support personnel to meet students' learning, professional, and emotional needs. However, serendipitously, these trends are in alignment with the advent and widespread use of generative AI tools in the quickest uptake of a new technology in the past generation [19]. New IPAs like ChatGPT and Bard are already being integrated for different tasks and goals across K-16 education. With these new tools, the one-on-one model of education is becoming mainstream as AI technology democratizes access to personalized services such as tutoring, coaching, mentorship, and even therapy [20–21].

### 2.1 Personalized learning

These new use cases for generative AI finally provide a solution to the nearly four-decade old one-on-one tutoring conundrum in education. As outlined in Bloom's 2 Sigma Problem, when comparing the performance of students receiving different types of instruction – traditional classroom instruction, mastery learning, and one-to-one tutoring – students who had one-to-one tutoring, personalized to their individual needs, performed significantly better than those who received group instruction in a traditional classroom setting [22]. The difference in performance between the two groups was approximately two standard deviations (or two sigma), which corresponds to a significant improvement in learning outcomes, which raised a critical question in the field of education: how can the effectiveness of one-to-one tutoring be replicated in group instruction settings, given the limited resources and practical constraints that most schools and educational institutions face?

Since Bloom's research, educators and researchers have explored various strategies to address the issue of individualized tutoring, such as differentiated instruction, adaptive learning technologies, and the incorporation of active learning techniques [23–25]. While learning outcomes were improved in these examples, addressing the challenge of scaling within a group instruction model and addressing issues of cost of instruction persists, until now. Generative AI and other advanced technologies have emerged as promising tools to help bridge the gap, offering more personalized and accessible educational experiences for students in group settings [26]. Therefore, by addressing Bloom's 2 Sigma Problem, AI can serve as a live tutor for anyone, while human experts can now provide additional in-depth knowledge as subject-matter experts and emotional support to learners [27]. Solutions like Numerade's AI tutor, Ace, were already able to generate customized study plans based on students' skill levels using a number of popular textbooks (https://www.numerade.com/). Trumbore [28] has since argued that lessons from the development of early computer-assisted instruction (CAI) systems, such PLATO, and intelligent AI tutors like Ace, offer insights into how digital assistants like ChatGPT can now be used as a tutor for students at all levels of instruction.

In reviewing what CAI like PLATO has been able to do in delivering individualized and self-paced instruction to learners in classroom settings, outlining the learning design use cases for the latest educational avatars is possible. For instance, through the use of computers instructional material was automatically presented to learners, learning progress monitored, and users were directed to resources or feedback based on their real-time needs. With CAI, learners are able to work independently and progress at their own pace, while receiving immediate feedback on their understanding of the material [29]. If a student's answer is incorrect, the program prompts them to review the material and make another choice, much like the functionality built into electronic textbooks in the early 2000s by Cengage and McGraw Hill [30].

Large language models and digital assistants now have the potential to serve in this role as personal tutors, offering individualized instruction to a broad range of learners [31]. By prompting students to revise and elaborate on their work, digital assistants can enhance learning outcomes, with their capacity for intricate dialog and comprehensive information delivery making them suitable tutors for students across various proficiency levels [32]. Moreover, once formative instruction and assessment can be largely automated, the time-constraints of educators become less restrictive, and more time can be spent on mentorship and apprenticeship in classroom and co-curricular settings [20].

Personalized learning is becoming increasingly attainable, as AI facilitates the customization of diverse educational elements, such as instructional methods, content formats, and curricula. AI-powered software can precisely evaluate students' abilities and knowledge deficiencies, consequently refining content to maximize engagement. Catering to a heterogeneous learner population, AI addresses the needs of advanced students, those grappling with particular concepts, introverted individuals, and those with distinct learning necessities. This transformation heralds a more inclusive and efficacious educational experience for all but will require a paradigm shift in how different subjects are taught and the student-teacher relationship.

## 2.2 Adapting to change

With the rapid advances we are seeing in the fields of AI, ML, and NLP, a new generation of AI-first tools will emerge for both teachers and students, as these groups have historically been early adopters of productivity software [33]. As such, as these digital assistants and educational avatars become more human-like with improved intelligence, both parties are likely to adopt chat-based conversational interfaces. Teachers, often overworked and underfunded, can benefit from AI tools that reduce workloads by drafting lesson plans, rubrics, assessments, syllabi, and more, thus allowing them to focus more on individual student attention. Meanwhile, students are constantly seeking creative ways to save time and gain advantages, which can be seen with the embracing of previous AI-driven resources like Photomath (https://photomath.com/en) to solve complex mathematical problems and Numerade to summarize the most salient points in a textbook's chapter [34–35]. The rapid spread of popular products in college environments through word of mouth further contributes to the adoption of these AI-first tools in education.

The broad adoption of these new emerging technologies will be accompanied by a major disruption in the educational status quo, especially considering how course materials are delivered, assessed, and even how credentialing will work. In fact, current pedagogical strategies, especially in college, prioritize the synthesis of existing information assessed through standardized testing or essay writing. The new capabilities of new generative AI can now automate the process and deliver nearly instantaneous results on nearly any topic, rendering previous instructional and assessment methods obsolete. Therefore, in order to accommodate these advancements, adjustments to classroom practices and assessment methods are necessary, similar to the introduction of the first electronic hand-held calculators (1967), the first commercially available personal laptop (1985), the broad accessibility of the internet (1996), the Google search engine (1998), and Wikipedia (2001) [36]. The development of next-generation tools for assessing student learning outcomes and awarding credentials, along with AI-leveraging tools, is on the horizon [37].

One of the most important elements in the paradigm shift for education is the growing importance of skills associated with critical thinking, digital and information fluency. Given how large language models process and generate content, fact-checking will become a central skill in the years ahead and one of the most important one's students should be equipped with after graduation [31]. As these AI technologies proliferate, concerns surrounding factual accuracy and potential algorithmic biases resulting from mismanaged or mislabeled datasets become more critical. Since algorithms are trained on data subject to human input, selection, judgment, and biases, these distortive effects can be rapidly and exponentially amplified in AI-generated content.

Fact-checking will be essential to counter AI-generated misinformation, as studies show people often find AI-composed content credible despite inaccuracies [38]. Trust in user-generated content and non-branded sources may decline, while blind trust in known personalities and experts might increase [39]. Additionally, the growing abstraction of complex systems, like web development, risks creating a generation competent in using tools without a deep understanding of underlying details [40]. The growing ease with which generative AI provides open access to a non-specialist audience, challenges in handling edge cases and crises that require detailed knowledge for effective problem-solving will potentially increase in tandem [41]. Educating students on the abilities, limitations, and responsible uses of these increasingly powerful content generators will be at the forefront of discussions on integration in the classroom moving forward.

Regardless of potential pitfalls, the use of AI tools, such as digital assistants and generative technologies, will be instrumental in preparing students for future success as new jobs are already being created. At the same time, the enthusiasm surrounding the quick adoption of these AI advancements can be ascribed to user-driven innovation. In this context, users significantly contribute to innovation by actively engaging with emerging technologies, as exemplified by ChatGPT. The lead-user theory aids in pinpointing users who are at the forefront of crucial trends and whose innovations possess a high probability of success. Such users play a vital role in fostering the engagement and adoption of novel technologies [42]. But that is not the only reason chatbots have seen widespread adoption.

The challenge of information overload in an increasingly complex world necessitates collective sense-making. As apprehensions regarding information management escalate, there is a willingness to embrace technologies capable of synthesizing and interpreting diverse information sources [43]. The fervor surrounding ChatGPT and analogous AI tools underscores the need to address these shared concerns while recognizing how broader cultural mechanisms shape user needs and expectations. Involving users in the innovation process via co-creation and lead-user participation is crucial for the successful adoption of emerging technologies, ultimately rendering AI an indispensable component in the future of work.

As such, there have been a plethora of examples and use cases of how to integrate these automated assistants in the classroom as of late. Integrating digital learning companions into educational settings has the potential to greatly enhance various aspects of teaching, including lesson planning, updating existing materials, and improvisation during class. By using chatbots as a collaborative tool, educators can streamline lesson planning, ensure consistency with curriculum goals, and make time to foster discussions about diversity and differentiation. The human-AI teaching and learning model can also be employed to refresh old teaching materials, aligning them with current values and pedagogy, and transforming traditional resources into more engaging activities. Furthermore, these assistants can serve as an improvisational aid, helping teachers to adapt to students' needs in real-time by offering novel explanations, essay topics, or scenarios.

Developing alongside these new integrations of generative AI into education are new jobs to be filled and a new skillset to prepare students for them. The rise of AI chatbots like ChatGPT and Microsoft's Bing AI have prompted companies to maintain up-to-date AI models to prevent hallucinations, misinformation, and unsettling threats that may be generated in their current state. This development has led to the emergence of a new professional skill called "prompt engineering," which refines chatbots' abilities to provide relevant and trustworthy responses through plain text interactions. However, experts are divided on whether prompt engineering constitutes a reliable, long-term career option or is merely a transient trend. The effectiveness of prompt engineering remains a subject of debate due to the inherent unpredictability of AI chatbot responses [6]. Regardless, the skill will be highly desirable moving forward. But exactly how does one develop the skill to become a so-called "prompt engineer"?

## 3. Metacognition

*The purpose of teaching is to inspire the desire for learning in them and make them able to think, understand, and question-* Richard Feynman.

Feynman, a Nobel Prize-winning physicist, defined science in the 1970s as "the belief in the ignorance of experts" [44]. The statement is truer now than ever and will be central to how we train students to critically evaluate AI-generated content moving forward. The ideas of Feynman reiterate the idea that the true objective of education is to cultivate metacognitive skills in students, rather than merely focusing on memorization and regurgitation of information in high-pressure academic environments. Metacognition, a term of significant importance within the realm of education and cognitive psychology, refers to an individual's awareness and understanding of their own thought processes, cognitive abilities, and learning strategies [45]. In essence, metacognition is the process of "thinking about thinking," enabling individuals to monitor, evaluate, and regulate their cognitive performance. By fostering metacognitive skills, learners are better equipped to identify their strengths and weaknesses, select effective learning strategies, set goals, and adjust their approaches as necessary to optimize learning outcomes [46]. This section discusses strategies for developing metacognition in students, while also acknowledging the irony in the role played by AI in the process. These AI tutors can be used to help develop a student's ability to learn how to learn, but educators should be doing so in order to equip students with the ability to question output by the same AIs.

Several strategies have been adopted to help students develop critical thinking and higher-level metacognitive skills. One approach would be constrained choice exercises. Constrained choice activities are an effective method for engaging higher education students in critical thinking and real-world decision-making skills development [47]. These activities involve carefully designed decision-making questions that limit students' inquiry scope, prompting them to apply abstract, theoretical knowledge to practical, real-world problems. By focusing on the most relevant concepts and encouraging students to make difficult, nuanced judgments, constrained choice activities facilitate the development of valuable workplace skills.

The structure of constrained choice activities typically includes a situation description, information for analysis, and a prompt for action or decision. The use of superlatives is essential in framing these specific decisions, forcing students to carefully analyze and decide on the best course of action. The simplicity and comparability of the resulting products of students' thinking enable easy reporting and focused classroom discussions. The benefits of constrained choice activities include promoting critical thinking, allowing for the addition of complexities or constraints during facilitation, and fostering productive classroom discussions [48]. This approach can be easily integrated into existing learning activities, requiring only subtle adjustments to activity prompts, thereby leading to improved student outcomes.

In addition to constrained choice, reflection activities also promote the development of metacognitive skills. The significance of reflection in the learning process is widely recognized, as it involves revisiting and reevaluating past experiences [48]. Reflection, simply defined as rethinking an event, is an inherent part of learning that does not always require deliberate encouragement. Factors such as perspective, mindset, background knowledge, emotional well-being, setting, and social context can influence not only how we perceive an event but also which aspects we experience or remember. For instance, watching a movie for the first-time results in a different cognitive experience compared to watching it a second time, even though the story and the observer remain the same.

Metacognitive prompts can be employed to promote reflection in the learning process. For instance, activities such as producing metacognitive journals offer students opportunities to ask and refine questions, analyze their thinking patterns, and recognize cognitive blind spots [49]. The hierarchical ordering of cognitive skills in Bloom's revised taxonomy can also be utilized to encourage students to reflect on past learning experiences in relation to their thinking and cognitive behaviors [50].

The following metacognitive prompts, based on Bloom's revised taxonomy, can help students reflect on their learning:

1. Can I identify patterns in my actions?

2. Were the strategies and skills I employed effective for this task?

3. How did my mindset influence my approach to the work?

4. Was my communication with others effective before, during, or after the learning process?

5. What have I learned about my strengths and areas for improvement?

6. How am I progressing as a learner?

7. What actions should I take next?

8. How can I best utilize my strengths for learning?

9. What steps should I take or resources should I use to address my challenges?

10. How can my learning environment be improved?

These prompts can aid students in reflecting on their learning experiences, particularly concerning their cognitive behaviors and thought processes. The integration of chatbots in this exercise is an additional way to promote metacognitive development. Having the AI interrogate metacognitive journaling processes adds an additional layer of dialogical engagement for student reflection.

Educators can also effectively utilize digital assistants to enrich classroom experiences and encourage metacognitive reflection among students. By incorporating AI into the daily learning process, teachers can create personalized and captivating learning environments. Integrating these chatbots into the classroom can be achieved through various means, including providing individualized feedback based on students' responses to metacognitive prompts, which allows them to reflect on their learning experiences and address their unique needs and areas of improvement. Additionally, these tools can generate tailored learning resources, like reading materials or quizzes, that are relevant and challenging for students based on their metacognitive reflections, learning level and style. They can also facilitate group discussions by generating thought-provoking questions or prompts, encouraging students to collaborate and learn from one another's experiences. Furthermore, educators can use digital assistants to monitor students' progress over time by analyzing their responses to metacognitive prompts, enabling them to adjust teaching strategies and provide timely support. Lastly, these new tools can aid educators in their own metacognitive reflection and professional growth by generating prompts or questions that help them evaluate their teaching practices, classroom environments, and student engagement.

## 4. Applied learning

Once students can identify the best way to learn through metacognitive exercises through restrained choice and/or reflective exercises, engaging in project-based and real-world learning should follow. Therefore, the full realization of the co-collaborative learning model AI now represents includes integrating applied learning theory and strategies in support of and to further develop metacognition in the curriculum. Applied learning is an educational approach that emphasizes the relevance and practical application of acquired knowledge and skills in real-world contexts [51]. The approach seeks to bridge the gap between theoretical understanding and practical implementation, fostering the development of problem-solving, critical thinking, and collaboration skills. Applied learning experiences can include case studies, internships, project-based learning, and other opportunities that actively engage learners in applying their knowledge to authentic situations [52–53]. With more professions expected to be proficient in AI tools, these authentic situations can be easily replicated in the classroom in scenario-based exercises.

Moreover, applied learning will become increasingly important as project-based and authentic assessment will become dominant given the user-friendly abilities of generative AI. As large amounts of content can be synthesized from existing datasets and automatically output, the importance of applied learning becomes even more pronounced. Moreover, as AI technologies advance and increasingly infiltrate various aspects of daily life, including education, learners will need to develop the ability to effectively utilize these tools in real-world settings. By integrating AI into applied learning experiences, students gain a deeper understanding of the technology's capabilities, limitations, and ethical considerations. In the context of generative AI, applied learning can help students develop essential competencies, such as critical analysis of AI-generated content, collaboration with AI tools for problem-solving, and responsible and ethical use of AI in their work. Moreover, applied learning experiences that incorporate AI can foster creativity, adaptability, and resilience in learners, preparing them for a future where AI will play a significant role in various professional and personal domains.

In order to operationalize this new approach to education that foregrounds metacognition through applied experiences, a paradigm shift in writing instruction is necessitated. In composition coursework, the primary focus has traditionally been on teaching grammar, syntax, and sentence-level writing, which underscores the significance of producing clear, concise, and grammatically accurate sentences [9]. The instruction of sentence-level writing commences with acquainting students with the fundamental elements of a sentence, subsequently familiarizing them with various parts of speech and their effective utilization. After mastering the basics, learners can concentrate on more sophisticated sentence-level writing abilities, such as syntax, sentence structure, and rhetorical devices. To reinforce these skills, students often engage in diverse writing assignments that emphasize crafting clear, concise, and grammatically sound sentences. The final stage of this iterative process involves feedback, which is typically provided by peers or instructors [10].

Nonetheless, with the advent of chatbots and digital assistants capable of generating drafts on various subjects given appropriate prompts, the pedagogical emphasis ought to transition towards cultivating students as editors, as opposed to sentence constructors. This paradigm shift has become evident in recent years, with individuals and even trained professionals increasingly relinquishing their roles as content creators. Presently, AI-powered tools have evolved to generate intricate written content. Consequently, these advanced digital assistants are tasked with producing sentence-level content, yielding initial drafts that humans will then refine through editing. In response to this human-AI collaborative dynamic, education, particularly at the postsecondary level, must restructure its approach to prepare students for their emerging roles as editors and evaluators of information.

Instead of the current pedagogic focus on sentence-level writing, educators should emphasize editing and prompt design and refinement as vehicles for enhancing metacognition. While the reframing sounds simple enough, the approach actually requires a reconceptualization of the problem which encompasses high-level thinking about the audience, contextual factors, and generated output. Students must cultivate language skills that enable effective interaction with AI and facilitate learning from AI-generated content. One potential criticism of this approach is that the editorial role does not foster sentence-level writing skills [54]. However, a novel model for writing education can be envisioned wherein students receive tutoring while editing. This model has the potential to be more efficacious in educational settings where students require guidance and instruction to refine their writing abilities.

## 5. Human-AI collaboration

Digital assistants have the potential to promote productivity and understanding of emerging technologies simultaneously. The potential to apply across various domains is already demonstrable, but an active approach that involves inquiry, co-creation, and collaboration needs to be adopted. Users must be engaged with the technology and strive to understand its capabilities and limitations to maximize its benefits. As such, this section will offer several practice strategies and assignments for educators and those interested in using generative AI tools to improve their metacognitive skills and learn how to critically evaluate a wide variety of content. The authors advise encouraging curiosity on both the part of the educator and/or student, and dedicate time to explore and question AI tools. Run the same question multiple times and pose the same question in various ways to obtain different perspectives. Always identify when work has been assisted by an AI tool, and do not accept or act on the first results without further investigation, evaluation and scrutiny. Seek to discover questions that produce both surprising and expected results, and learn to explain and defend why one accepts or prefers certain outputs over others.

As noted, prompt engineering, which involves generating prompts or assignments using AI, can be a useful tool in developing essential metacognitive skills in students. However, the evaluation of the output from prompt engineering requires another metacognitive level of thinking, where students must critically analyze and evaluate their work and question their input to revise and determine what and/or how output is working. Therefore, it is crucial to teach students two different skills: first, how to write a paper about a specific topic, and second, how to think about early stages of a project, prompt, or assignment. This second skill is not narrow and becomes a metacognitive skill in and of itself. An example of an assignment that may be considered in developing these skills would involve identifying a topic, generating an initial response from a chatbot, and then iteratively evaluating and refining that as follows using a chatbot of choice (**Table 1**).

| |
|---|
| **Assignment Title:** Investigating Metacognition Development through AI-Assisted Editing |
| **Objective:** To explore the development of metacognition in students by utilizing chatbots in the process of editing and refining AI-generated content |
| **Instructions:**<br>1. Choose a topic of interest within your field of study. Develop a clear, concise prompt related to your chosen topic that will guide the chatbot in generating an initial draft.<br>2. Use the chatbot to generate an initial 250-word essay section based on the prompt you created. Save the AI-generated text as a separate document.<br>3. Review the 250-word section and prompt the chatbot for several variations by varying the prompt. Compare the results and feel free to change the essay from your initial plan.<br>4. Ask the chatbot to complete the essay with an additional 250-word section. Save the text as a full (short) essay.<br>5. Carefully read the AI-generated essay, noting any inaccuracies, unclear phrasing, or overly general language. As you read, consider the following questions (for now, do not worry about sentence-level issues):<br>• Where does the AI writing require more specific examples to support general claims?<br>• Where do you see strong factual claims that need to be checked for accuracy?<br>• Most importantly, where do you see interesting directions for inquiry proposed by the AI but which you did not initially consider?<br>6. Engage in a dialog with the chatbot to understand the rationale behind specific choices in the AI-generated content. Ask the AI questions about its choices, and use the feedback to refine your editing process further.<br>7. Continue the iterative process of editing and consulting the chatbot until you have a polished final draft. Ensure that you maintain ultimate decision-making authority over the final edited product.<br>8. Write a 300-word reflection on your experience using the chatbot as a tutor and editor. Discuss how the process influenced your metacognitive thinking and editing skills. Address the following questions in your reflection: |

- How did interacting with the chatbot impact your understanding of the topic and the structure of the essay?
- In what ways did the AI-assisted editing process challenge or support your critical thinking and reading skills?
- How did engaging with the chatbot as a tutor influence your sentence-level writing and editing abilities?

9. Submit the polished final draft of your essay, the original AI-generated text, and your reflection.

**Grading Criteria:**
- Clarity and coherence of the final edited essay
- Effectiveness of AI-generated content in addressing the prompt
- Quality of editing, including grammar, syntax, and overall structure
- Depth of reflection on the AI-assisted editing process and its impact on metacognitive thinking and editing skills
- Engagement with the chatbot and the extent to which feedback was incorporated into the editing process

**Table 1.**
*Investigating metacognition development through AI-assisted editing assignment.*

In the AI-assisted editing assignment outlined above, the learning outcomes and metacognitive value for students are multifaceted. By actively involving students in the process of refining AI-generated content, this approach promotes a deeper understanding of the subject matter, enhances editing skills, and fosters critical thinking.

The primary learning outcomes of this approach include:

1. Development of critical reading and analysis skills, as students must evaluate the quality and coherence of the AI-generated content, ensuring it aligns with the intended argument or message.

2. Improvement of sentence-level writing and editing abilities, as students engage in an iterative process of refining the AI-generated text, addressing any inaccuracies, unclear phrasing, or grammatical errors.

3. Enhancement of collaboration and communication skills, as students interact with the AI to understand the rationale behind specific content choices, refining their editing process based on the feedback received.

The metacognitive value of this approach lies in the following aspects:

1. Fostering self-awareness of one's own thought process and understanding, as students must consider the purpose and structure of the AI-generated content, reflecting on how it contributes to the overall argument or message.

2. Encouraging students to engage in higher-order thinking, as they must consider the intended audience, contextual factors, and the generated output while editing the AI-generated content.

3. Promoting self-regulation and autonomy, as students maintain ultimate decision-making authority over the final edited product, guiding the AI-generated content to align with their understanding and objectives.

The AI-assisted editing approach not only enhances students' writing and editing skills but also fosters metacognitive development by encouraging critical analysis, self-awareness, and self-regulation.

Supplementary assignments and activities can enhance student engagement in the learning process by actively involving them in the creation and evaluation of educational materials. For example, when teaching the combining of independent clauses into longer sentences, rather than merely supplying students with pre-made worksheets, instructors can employ AI tools to generate five distinct worksheets, allowing students to attempt each and request a grade upon completion. This method can serve as a heuristic model for education, as it encourages students to formulate questions they need to address (see **Table 2**) [55]. However, one challenge with this approach is that, while students may be adept at reading articles and collecting information, they might encounter difficulties in identifying contentious issues or debates within the subject matter [56].

To help students identify controversies in a field more effectively, teachers can encourage them to develop their own stylistic preferences. This approach allows students to learn grammar more efficiently, freeing up time for them to concentrate on higher-level metacognitive skills. For example, in computer science courses,

| |
|---|
| **Assignment Title:** Heuristic Learning with AI-Generated Sentence-Combining Worksheets |
| **Objective:** To engage students in a heuristic approach to learning the combining of independent clauses using AI-generated worksheets, promoting active learning and self-assessment through an iterative process. |
| **Instructions:**<br>1. The teacher will use an AI tool to create five different worksheets teaching sentence combining, each with varying levels of complexity.<br>2. Students will select one worksheet to start and complete the exercises, keeping track of the time taken and their level of confidence in their answers.<br>3. Upon completing the first worksheet, students will request a grade from the AI tool and review the feedback provided.<br>4. Based on the feedback, students will identify areas of difficulty or uncertainty in independent clause combination and generate a list of questions to guide their learning.<br>5. Students will conduct independent research on the identified concepts, using articles and other resources to deepen their understanding of independent clause structure.<br>6. After the research phase, students will select another AI-generated worksheet, complete the exercises, and ask for a grade again. They will repeat this process until they have worked through all five worksheets.<br>7. Following the completion of all five worksheets, students will write a 250-word reflection on their learning process, addressing the following questions:<br>• How did the heuristic approach impact your understanding of sentence-combining concepts?<br>• What challenges did you face in identifying the variation in stylistic effect created by different combinations, and how did you overcome them?<br>• How did engaging with AI-generated worksheets and self-assessment support your learning and retention of independent clause structure? |
| **Grading Criteria:**<br>• Completion of all five AI-generated worksheets<br>• Quality and depth of the research conducted on sentence-combining concepts<br>• Clarity and coherence of the written reflection<br>• Engagement with the heuristic learning process and improvement in clause structure knowledge and skills |

**Table 2.**
*Heuristic learning with AI-generated sentence-combining worksheets assignment.*

students progress through various abstraction levels, initially learning to code before ascending the abstraction hierarchy [57]. A similar method can be applied to humanities courses (**Table 3**). In these courses, students can start by focusing on sentence-level design, gradually shifting their attention to the organization and structure of essays. Subsequently, they can concentrate on literature reviews and their integration within broader academic discussions. Utilizing a heuristic framework enables students to fluidly transition between abstraction layers, dedicating more time to higher-level metacognitive skills.

By incorporating AI for syntactic-level writing, students can further enhance their focus on advanced metacognitive skills. AI assistance allows them to generate initial drafts and engage in the editing process while exploring the different abstraction layers in their writing. Consequently, students can develop a deeper understanding of the subject matter, analyze their own thought processes, and hone their writing skills, all while leveraging AI technology to support their learning journey [58].

As such, the collaboration between humans and AI in education presents a unique opportunity to rethink the way we teach writing and related skills. While it remains essential to equip students with a robust foundation in grammar and syntax, it is crucial to strike a balance by emphasizing higher-level metacognitive skills, such as

| |
|---|
| **Assignment Title:** Abstraction Layers in Writing and Developing Metacognitive Skills |
| **Objective:** To facilitate students' understanding of higher-level metacognitive skills by guiding them through a series of abstraction layers in writing, from sentence-level design to broader literature analysis. |
| **Instructions:**<br>1. Sentence-level design: Students will create a set of stylistic preferences at the sentence-level based on the course material, providing a foundation for quick grammar learning and enabling them to focus on higher-level metacognitive skills.<br>2. Organization and essay structure: Students will write a 500-word essay on a topic relevant to the course. They must apply the stylistic effects created in step 1 and demonstrate a clear organization and structure in their essay.<br>3. Literature review and broader conversation: Students will conduct a literature review on the essay topic, identifying at least five key sources that contribute to the academic conversation around the issue. They will analyze the sources, outlining their main arguments and their relation to the chosen topic.<br>4. Integrating abstraction layers: Students will revise their essay from step 2, incorporating insights from the literature review in step 3. They will ensure that their essay engages with the broader conversation and demonstrates an understanding of the various abstraction layers in writing.<br>• How did the process of moving between abstraction layers impact your understanding of the topic and writing skills?<br>• In what ways did engaging with various abstraction layers contribute to the development of higher-level metacognitive skills?<br>• How did using AI for syntactic-level writing influence your focus on higher-level metacognitive skills?<br>5. Reflective analysis: Students will write a 300-word reflection on their experience moving between abstraction layers, addressing the following questions: |
| **Grading Criteria:**<br>• Quality and coherence of the stylistic effects created<br>• Clarity, organization, and structure of the essay and literature review<br>• Successful integration of abstraction layers in the revised essay<br>• Demonstrated understanding of higher-level metacognitive skills in the reflective analysis<br>• Engagement with AI tools for syntactic-level writing and its impact on the learning process |

**Table 3.**
*Abstraction layers in writing and developing metacognitive skills assignment.*

analysis and evaluation. By shifting the instructional focus from sentence generation to editing, we can leverage AI's capabilities to generate initial drafts while humans refine and polish them. This novel approach not only fosters a deeper understanding of the writing process but also cultivates critical thinking and metacognitive skills in students. Ultimately, the integration of AI into the educational landscape holds the potential to transform our learning and teaching methods, paving the way for a more individualized and effective educational experience for learners from all walks of life.

## 6. Conclusion

The integration of intelligent virtual assistants (IVAs) in education has the potential to revolutionize the learning process and equip learners with the necessary skills for the future workforce. To meet this growing demand, specific exercises, assignments, assessments, and even classes and certificates will be developed to address the need for prompt engineering upskilling. The incorporation of metacognitive skills into curricula across all disciplines will become a key pedagogical strategy, enabling students to develop evaluative skills crucial for success in the future. The use of tailored AI tools for different learning outcomes will also enhance the learning experience, allowing students to generate quizzes and gamify their learning experience, leading to increased efficiency in education. The acquisition of advanced metacognitive skills will become a core requirement for the upcoming generation of learners, much like typing was a technical skill taught in the past.

In the suggested heuristic model for education, AI generates initial drafts for papers, allowing students to engage in dialogs with the AI to comprehend the reasoning behind specific choices. This iterative interaction fosters both editorial and sentence-level writing skills, with students retaining control over the final edited product and benefiting from the AI's guidance. Implementing this pedagogical approach involves students initiating the process with AI-generated text, using it as a tutoring resource, and alternating between editing, verification, idea generation, and evaluation. This method encourages critical analysis and reading skills among students. The integration of AI as both a tutor and editor in this innovative educational model enhances metacognitive thinking, advances sentence-level writing, and offers customized feedback and support. By reorienting writing course objectives and adopting AI technologies, educators can transform writing instruction, equipping students for future workforce demands.

Moreover, the rise and proliferation of generative AI technology heralds a promising future for higher education by offering personalized learning experiences and content generation capabilities. AI-powered virtual assistants can provide tailored feedback and support to students, aiding them in grasping intricate concepts and improving their academic performance. However, it is crucial to acknowledge that AI technology cannot replace the human interaction and critical thinking skills nurtured through conventional college education. Therefore, the future of college education is likely to encompass a blend of AI technology and traditional classroom instruction, with AI complementing and supplementing conventional teaching methods. This blended approach allows students to benefit from AI's advantages while still acquiring valuable human interaction and critical thinking skills indispensable for success in the workforce. Consequently, it is essential to persist in exploring AI technology's potential in education and optimize its utilization to enhance learning outcomes for upcoming generations of learners.

As AI continues to evolve and advances are made daily in the field, driven by increased processing power and more accessible hardware to run supercomputers, its implications for education remain a moving target. The rapid progress of AI models, such as GPT-4, in various contexts highlights the importance of determining whether prompt design and engineering will become a skill students need to learn. The contrasting experiences of start-ups and users of different AI models underscore the importance of ongoing research and adaptation to fully harness AI's potential in education [59]. Ultimately, by embracing human-AI collaboration and reimagining applied learning to support metacognition, we can transform educational experiences, empower learners, and prepare them for the dynamic demands of the future workforce.

## Conflict of interest

The authors declare no conflict of interest.

## Author details

James Hutson* and Daniel Plate
Lindenwood University, Saint Charles, USA

*Address all correspondence to: jhutson@lindenwood.edu

## References

[1] Tulshan AS, Dhage SN. Survey on virtual assistant: Google assistant, siri, cortana, alexa. In: Advances in Signal Processing and Intelligent Recognition Systems: 4th International Symposium SIRS 2018, Bangalore, India, September 19–22, 2018, Revised Selected Papers 4. Singapore: Springer; 2019. pp. 190-201

[2] Chow AR. How ChatGPT Managed to Grow Faster than TikTok or Instagram. Time. 2023. Available from: https://time.com/6253615/chatgpt-fastest-growing/

[3] Harrison M. Economists Say that AI is a Doomed Bubble. Futurism. 2023. Available from: https://futurism.com/economist-ai-doomed-bubble

[4] Williams T. Some Companies are Already Replacing Workers with ChatGPT, Despite Warnings it Shouldn't be Relied on for 'anything important.' Fortune. 2023. Available from: https://fortune-com.cdn.ampproject.org/c/s/fortune.com/2023/02/25/companies-replacing-workers-chatgpt-ai/amp/

[5] Kim K, de Melo CM, Norouzi N, Bruder G, Welch GF. Reducing task load with an embodied intelligent virtual assistant for improved performance in collaborative decision making. In: 2020 IEEE Conference on Virtual Reality and 3D User Interfaces (VR). New York: IEEE; 2020. pp. 529-538

[6] Tangermann V. Get a Load of This New Job: "Prompt Engineers" Who Act as Psychologists to AI Chatbots. Futurism. 2023. Available from: https://futurism.com/prompt-engineers-ai

[7] Clark D. PedAIgogy – New Era of Knowledge and Learning Where AI changes Everything. Donald Clark Plan B. 2023. Available from: http://donaldclarkplanb.blogspot.com/2023/03/pedaigogy-new-era-of-knowledge-and.html?m=1

[8] Pal K. Evaluation of a scenario-based Socratic style of teaching and learning practice. In: Enhancing Teaching and Learning with Socratic Educational Strategies: Emerging Research and Opportunities. Hershey, Pennsylvania: IGI Global; 2022. pp. 121-144

[9] Foltz-Gray D. Responses to error: Sentence-level error and the teacher of basic writing. Research and Teaching in Developmental Education. 2012;**28**(2):18-29

[10] Hesse D. We know what works in teaching composition. The Chronicle of Higher Education. 2017;**3**. Available from: https://www.chronicle.com/article/we-know-what-works-in-teaching-composition/

[11] Ma WY. Democratizing content creation and dissemination through AI technology. Proceedings of The Web Conference. 2020;**2020**:3143-3143

[12] 2023 Trends in Higher Education. Hanover Research. 2023. Available from: https://insights.hanoverresearch.com/hubfs/2023-Trends-in-Higher-Education.pdf?li_fat_id=1e69f6c5-1686-4335-8451-72c8d5fbc45b

[13] Shulock N, Offenstein J. Career Opportunities: Career Technical Education and the College Completion Agenda. Part I: Structure and Funding of Career Technical Education in the California Community Colleges. Washington, D.C.: Institute for Higher Education Leadership & Policy; 2012

[14] Bishop MM. Addressing the Employment Challenge: The Use of

Postsecondary Noncredit Training in Skills Development. Washington, D.C.: American Enterprise Institute; 2019

[15] Jones T, Ramirez-Mendoza J, Jackson V. A Promise Worth Keeping: An Updated Equity-Driven Framework for Free College Programs. Washington, D.C.: Education Trust; 2020

[16] Counselman Carpenter EA, Meltzer A, Marquart M. Best practices for inclusivity of deaf/deaf/hard of hearing students in the synchronous online classroom. World Journal of Education. 2020;**10**(4):26-34

[17] Braumberger E. Library services for autistic students in academic libraries: A literature review. Pathfinder: A Canadian Journal for Information Science Students and Early Career Professionals. 2021;**2**(2):86-99

[18] Wingard D. Data-driven automated decision-making in assessing employee performance and productivity: Designing and implementing workforce metrics and analytics. Psychosociological Issues in Human Resource Management. 2019;7(2):13-18

[19] Skates A. Five Predictions for the Future of Learning in the Age of AI. Everyday AI. 2023. Available from: https://a16z.com/2023/02/08/the-future-of-learning-education-knowledge-in-the-age-of-ai/?utm_campaign=GSVN2K&utm_medium=email&_hsmi=245282162&_hsenc=p2ANqtz-_Ekqx7DSmPF7JCn_1EVprjfxEDqq6NGJJ8UZuiQaCOdKKEbZF2ibPx6WIbF8Cg83kLd9WscK_J_8Wtbh2iONIWc3o_Zw&utm_content=245282162&utm_source=hs_email

[20] Hernandez PR, Ferguson CF, Pedersen R, Richards-Babb M, Quedado K, Shook NJ. Research apprenticeship training promotes faculty-student psychological similarity and high-quality mentoring: A longitudinal quasi-experiment. Mentoring & Tutoring: Partnership in Learning. 2023;**1**:1-21

[21] Irby BJ, Pashmforoosh R, Lara-Alecio R, Tong F, Etchells MJ, Rodriguez L. Virtual mentoring and coaching through virtual professional leadership learning communities for school leaders: A mixed-methods study. Mentoring & Tutoring: Partnership in Learning. 2023;**11**(3):1-33

[22] Bloom BS. The 2 sigma problem: The search for methods of group instruction as effective as one-to-one tutoring. Educational Researcher. 1984;**13**(6):4-16

[23] Mulwa C, Lawless S, Sharp M, Arnedillo-Sanchez I, Wade V. Adaptive educational hypermedia systems in technology enhanced learning: a literature review. In: Proceedings of the 2010 ACM conference on Information technology education. 2010. pp. 73-84

[24] Alamri HA, Watson S, Watson W. Learning technology models that support personalization within blended learning environments in higher education. TechTrends. 2021;**65**:62-78

[25] Dejene W. The practice of modularized curriculum in higher education institution: Active learning and continuous assessment in focus. Cogent Education. 2019;**6**(1):1-16. Research-Article

[26] Mallik S, Gangopadhyay A. Proactive and reactive engagement of artificial intelligence methods for education: A review. arXiv preprint arXiv:2301.10231. 2023. DOI: 10.48550/arXiv.2301.10231

[27] Hu YH, Fu JS, Yeh HC. Developing an early-warning system through robotic

process automation: Are intelligent tutoring robots as effective as human teachers? Interactive Learning Environments. 2023;**1**:1-4

[28] Trumbore A. ChatGPT could be an effective and affordable tutor. The Conversation. 2023. Available from: https://theconversation-com.cdn.ampproject.org/c/s/theconversation.com/amp/chatgpt-could-be-an-effective-and-affordable-tutor-198062

[29] Cope B, Kalantzis M. A little history of e-learning: Finding new ways to learn in the PLATO computer education system, 1959–1976. History of Education. 2023:1-32. DOI: 10.1080/0046760X.2022.2141353

[30] Brown BW. Vision and reality in electronic textbooks: What publishers need to do to survive. Educational Technology. 2012:30-33. Available from: https://www.bgsu.edu/news/2013/09/vision-and-reality-in-electronic-textbooks.html

[31] Kasneci E, Seßler K, Küchemann S, Bannert M, Dementieva D, Fischer F, et al. ChatGPT for good? On opportunities and challenges of large language models for education. Learning and Individual Differences. 2023;**103**:102274

[32] Pedro F, Subosa M, Rivas A, Valverde P. Artificial Intelligence in Education: Challenges and Opportunities for Sustainable Development. UNESCO; 2019

[33] García-Peñalvo FJ. The perception of artificial intelligence in educational contexts after the launch of ChatGPT: disruption or panic? Education in the Knowledge Society. 2023;**24**:1-9

[34] Peters VL, Hewitt J. An investigation of student practices in asynchronous computer conferencing courses. Computers & Education. 2010;**54**(4):951-961

[35] Mekni M. An artificial intelligence based virtual assistant using conversational agents. Journal of Software Engineering and Applications. 2021;**14**(9):455-473

[36] Valtonen T, López-Pernas S, Saqr M, Vartiainen H, Sointu ET, Tedre M. The nature and building blocks of educational technology research. Computers in Human Behavior. 2022;**128**:107123

[37] Nguyen LA, Habók A. Tools for assessing teacher digital literacy: A review. Journal of Computers in Education. 2023:1-42. DOI: 10.1007/s40692-022-00257-5

[38] Moon WK. Impactful Fact-Checking for Science: Communication Strategies Applying Artificial Intelligence as the Source of Information (Doctoral Dissertation). Austin, TX: The University of Texas at Austin; 2022

[39] Koivisto E, Mattila P. Extending the luxury experience to social media–user-generated content co-creation in a branded event. Journal of Business Research. 2020;**117**:570-578

[40] Van Beek TJ, Erden MS, Tomiyama T. Modular design of mechatronic systems with function modeling. Mechatronics. 2010;**20**(8):850-863

[41] Whitewood A. Introduction on AI approaches in capital markets. The AI Book: The Artificial Intelligence Handbook for Investors, Entrepreneurs and FinTech Visionaries. 2020. pp. 150-156

[42] Franke N, Von Hippel E, Schreier M. Finding commercially attractive user innovations: A test of lead-user theory. Journal of Product Innovation Management. 2006;**23**(4):301-315

[43] Roetzel PG. Information overload in the information age: A review of the literature from business administration, business psychology, and related disciplines with a bibliometric approach and framework development. Business Research. 2019;**12**(2):479-522

[44] Feynman RP, Cashman D. The Pleasure of Finding Things Out. Ashland, Oregon: Blackstone Audio, Incorporated; 2013

[45] Azevedo R. Reflections on the field of metacognition: Issues, challenges, and opportunities. Metacognition and Learning. 2020;**15**:91-98

[46] Hamzah H, Hamzah MI, Zulkifli H. Systematic literature review on the elements of metacognition-based higher order thinking skills (HOTS) teaching and learning modules. Sustainability. 2022;**14**(2):813

[47] Sibley J, Roberson B, O'Dwyer B. Constrained Choice Activities: A Simple Way to Improve Critical Thinking. Educause Review. 2023. Available from: https://er.educause.edu/articles/2023/1/constrained-choice-activities-a-simple-way-to-improve-critical-thinking

[48] Renshaw I, Chow JY. A constraint-led approach to sport and physical education pedagogy. Physical Education and Sport Pedagogy. 2019; **24**(2):103-116

[49] Walshe N, Driver P. Developing reflective trainee teacher practice with 360-degree video. Teaching and Teacher Education. 2019;**78**:97-105

[50] Alt D, Raichel N. Reflective journaling and metacognitive awareness: Insights from a longitudinal study in higher education. Reflective Practice. 2020;**21**(2):145-158

[51] Murtadho F. Metacognitive and critical thinking practices in developing EFL students' argumentative writing skills. Indonesian Journal of Applied Linguistics. 2021;**10**(3). DOI: 10.17509/ijal.v10i3.31752np

[52] Ash SL, Clayton PH. Generating, Deepening, and Documenting Learning: The Power of Critical Reflection in Applied Learning. 2009

[53] Ovenden-Hope T, Blandford S. Understanding Applied Learning: Developing Effective Practice to Support all Learners. Abingdon, Oxfordshire: Routledge; 2017

[54] Yates R, Kenkel J. Responding to sentence-level errors in writing. Journal of Second Language Writing. 2002;**11**(1): 29-47

[55] Vieira EA, SILVEIRA AC, Martins RX. Heuristic evaluation on usability of educational games: A systematic review. Informatics in Education. 2019;**18**(2):427-442

[56] Brecke R, Jensen J. Cooperative learning, responsibility, ambiguity, controversy and support in motivating students. Insight: A Collection of Faculty Scholarship. 2007;**2**:57-63

[57] Touretzky DS, Marghitu D, Ludi S, Bernstein D, Ni L. Accelerating K-12 computational thinking using scaffolding, staging, and abstraction. In: Proceeding of the 44th ACM Technical Symposium on Computer Science Education. 6 Mar 2013. pp. 609-614

[58] Clancey WJ. Heuristic classification. Artificial Intelligence. 1985;**27**(3):289-350

[59] Harsha N, King N, Mayer SMK, Carignan D, Horvitz E. Capabilities of GPT-4 on medical challenge problems. Microsoft OpenAI; 2023

Chapter 4

# AI of Public Communication for the Vulnerable: A Focus on Voice-Based Chatbots and Policy Suggestions

*Jinsoon Song*


## Abstract

The provision of voice-based chatbot public services by the government and public sectors is a means to realize digital inclusion in ICT society by ensuring information accessibility for the socially vulnerable. This paper presented five experts with information on artificial intelligence based on interviews on the relevance and utility, efficiency, accessibility, and openness of related technologies in providing public services and voice-based services. Moreover, a small online survey confirms that citizens' attitudes, perceptions, and expectations of public institutions ahead of voice-based service provision are positive. Then, technical aspects of the voice-based chatbot are discussed, defining its significance and necessity in public service, and the implications of the considerations of the use of voice-based services are drawn out. As an aspect of public communication for open government facilitating digital inclusion, the chatbot voice-based service in public service entails the practice of social values, affording the socially vulnerable opportunities to participate in intelligent government, to strengthen information accessibility, and to guarantee and strengthen human rights and basic rights.


**Keywords:** voice-chatbot, public communication, digital inclusion, the socially vulnerable, information accessibility, intelligence government

## 1. Introduction

Since the COVID-19 pandemic began in 2019, there has been widespread expansion and growth in activities and events within the virtual world. As society moves rapidly to digital, the social acceptance of digital communication technology is forming around the concept of the "personalized network" [1]. However, this vaguely encompasses the complex, sometimes conflicting values of personalization and networking. While personalization entails disintegration and diversification, networking entails aggregation and connection. The problem is that we need to think about how we are living in this contradictory situation of dismantling and reshaping, and whether this is proceeding adequately by embracing the socially vulnerable from the perspective of equity and equality. Government administrative services continue to innovate by combining technologies and networks developed by e-government for use in intelligent government in ICT society. Artificial intelligence (AI) technology is widely used in various industries to improve qualitative support for decision-making and problem-solving through the exploration of various types of mechanical intelligence, including natural language understanding, robotics, expert systems, neural networks, and machine learning [2]. The pandemic has also become a catalyst for business innovation: 76% of companies are planning long-term IT changes, and they are still introducing and providing various AI tech services such as big data, cloud, blockchain, IoT (Internet of Things), and open API. Due to recent AI advances and abundant data provision, government agencies have adopted chatbots and have begun to be applied to quite complex tasks in various fields such as laws, taxes, social welfare and civil complaints, and housing finance [3]. For example, among government tasks, AI applications are found to be suitable in six types: artificial intelligence resource allocation, large-scale data set sharing, lack of manpower, procedures and simple iterations, various data collection and summary, and future prediction [2, 4]. The use of AI apps in the public sector yields benefits such as reducing costs and work burdens, improving productivity, creating new employment opportunities, resolving resource allocation problems, providing public services, and enhancing civic satisfaction [5]. AI chatbots[1] are a representative citizen-friendly model, and securing various channels for public communication progress between the government and citizens for which chatbots serve as the introduction to administrative services as emotional secretaries, which are the advanced technologies, has great significance. It is possible to communicate with citizens 24 hours a day and provide personalized services, so the system is being quickly introduced and established in the government and public sectors. Our small survey on the convenience of such affordances as interworking with existing smart devices and mixing voice services found that citizens' attitudes are very open and expectations are high (see Section 3.2). In particular, the term "conversational commerce" is already being implemented in various ways through applications in business and industry, especially in the financial sector. For example, users do not have difficulty gaining access in real life to needed products through recommendations of products to suit their needs or tastes through chatbots [6]. Voice consultants such as Amazon's Alexa or Google Assistant can be used on designated devices or regular smartphones, making it available to numerous extended users. More than 28 million so-called "smart speakers" were reported worldwide in the third quarter of 2019 alone [7].

Conversational interfaces use natural language to interact with applications, which presupposes that they will have great potential for civil services and lower barriers to interaction with governments. The open government aims to focus on providing citizen-centered services in a form that easily accesses, utilizes, and shares information provided by the public sector, improves citizens' participation in public sector activities, and reflects their intentions. The main issues of open government are transparency, participation, and cooperation to create participatory and cooperative dialog [8]. Voice-based services using natural language serve to strengthen and protect the human and basic rights of social vulnerables and can function as public communication channels for realizing open government. Public communication between citizens and governments uses everyday language to enable government

[1] Machine learning algorithms, process automation, image recognition software, and voice recognition interfaces, including chatbots.

departments and institutions to provide, share, disclose, and utilize desired information smoothly and quickly, and engage their opinions in the policy-making process, enabling cooperation between governments and citizens [2]. The issue of the universality of digital technology, that is, citizens' access to digital technology, should be considered to maintain citizens' economic, geographic, and psychological equity in digital government. In particular, promoting the participation of social vulnerables is a key task of intelligent government that should be realized. This paper discusses how voice-based services using natural language can serve as a public communication channel for realizing open government by helping strengthen human rights and basic rights, digital inclusion, and the protection of information accessibilities by bridging the information gap of the socially vulnerable. This study is aimed at 1. determining the reasons for introducing voice service chatbots for the socially disadvantaged, 2. identifying the considerations for preparing a voice-based service platform for public sectors, 3. offering policy suggestions for the social vulnerable, and 4. drawing out the appropriate implications.

## 2. Voice chatbot usages in governments worldwide and the definition of the digital divide of the socially vulnerable

### 2.1 Global government status using chatbot's voice-based services

For many years, government agencies globally have provided voice-based chatbot services to provide rapid information and smooth communication from government to citizens (**Table 1**). It has been stated that this is to improve service delivery effectiveness and the efficiency of internal manpower management [10].

As shown in **Table 1** below, various public administrative services are provided by governments around the world. While seeking to improve civic services and government functions, the public sector is providing convenient and detailed automation services. Examples of chatbot usages in the public sector are as follows [4]: (1) answering citizens' questions, visits, and inquiries through an automated AI-based customer support system that performs such functions as representing life events, business events, and public services; (2) providing guidelines for searching documents (including legal documents) and completing forms for citizens; (3) receiving citizens' opinions; (4) translating government information; (5) drafting documents [2]; and (6) protecting security and predicting criminal activities [9]. Chatbots incorporate a "supervised learning" algorithm that allows them to continuously learn from interactions with humans and improve the accuracy of the responses they provide so that they actually become smarter. In addition, the "transparency bot" aims to increase transparency in government information development and provision by monitoring and reporting government actors' participation in Wikipedia on social media.

## 3. Definition of the concept of the digital divide for the vulnerable

Generally speaking, a vulnerable group is an individual or class that finds it difficult to maintain its current economic status when faced with social risks (young people in the job search after graduation, middle-aged people unable to get a job after early retirement), which are inevitable in the course of employment or economic

| | | |
|---|---|---|
| **Department of Homeland Security** | **EMMA** | **Guided and informed by U.S. Citizenship and Immigration Services (English Voice Service) the Mississippi State Government** |
| the Mississippi State Government | MISSI | - Enter concerns through voice input via Amazon's voice assistant Alexa<br>- Provide residents with information on places to visit and events such as public services and taxes, medical services, public transportation, family services, employment opportunities |
| San Francisco | PAIGE government | San Francisco's procurement chatbot application for internal workers keeps employees out of all the busy processes of government procurement. |
| Kansas City | OpenDataKC | Kansas City's open data portal is developed to help people find details about the city in one location. Interacting with people and providing immediate information to prevent people from wasting their time passing through the portal. |
| Government of Los Angeles | LACity Alexa | Launched LA City, a voice-enabled application for Amazon Alexa devices. Through this, the user may request information on events/news occurring in the city.<br>General administration |
| General administration | Mrs. Landingham | Guides new employees through a complete in-house onboarding process, such as filling out forms and organizing deliberations |
| Dubai Government | Rammas | - The first government chatbot application launched on the Google AI platform<br>- Provides the ability to process people's requests (claims and payments) 24 hours a day, process data, and make more accurate decisions |
| Government of Singapore | Gov.sg | Information about government agencies, news, press releases, people, and policies can be easily found and complaints about the negligence of all public services can be raised and tracked. Chatbots extract information from government-only portals already set up for public use. |
| the London Government | TravelBot | Bus arrival, route status, service update (bus/railway), and map information provided. |
| Australian Government | Alex | Australian IRS chatbots support mainly services related to taxation, property rights, income and deductions, filing declarations, and tax-related people and business issues. This service saves time and creates a better experience of navigating information by going directly to the content you are looking for. |
| Government of Bonn City, Germany | Botty Bonn | To help people with all urban problems and other administrative information. |
| India Maharashtra Government | PMC | - Provide registration, online services, tax returns, health issues, finances, and driver's license information.<br>- Chatbot applications provide up-to-date information on analytical data (e.g., health data) across all public services<br>- More transparency between government agencies and people. |

*Source: Song, [9], based on https://blog.vsoftconsulting.com/blog/15-governments-agencies-that-use-chatbots*

**Table 1.**
*Global government voice chatbot usage.*

| The stage of change | Introduction period | Taking a leap forward | The saturation period | The oversaturation period |
|---|---|---|---|---|
| Types of digital divide | access divide | usage divide | divide steamming from the quality of use | divide guaranteeing the fundamental human right. |
| Term | Initial information divide | First information divide | The 2nd information divide | The 3rd information divide. |
| Explanation | The difference between those who are accessible and those who are not. | The difference between a user and a non-user. | Differences between users and users. | The difference between the socially vulnerable and the non-vulnerable. |

*Source: Reconfigure Song, [9] reference to Min [15].*

**Table 2.**
*Definition of the concept of digital divide by period.*

activity or in the face of unexpected accidents (disease, industrial accidents, unemployment, etc.; [11]). Social minorities refer to those who have relatively limited opportunities for social participation compared to other classes due to economic, physical, and other conditions and are more likely to be excluded from opportunities to receive equal benefits as members of a society without state public intervention. They tend to be culturally alienated due to social, economic, and physical conditions, and thus form a socially vulnerable class in our society [12]. The class is more vulnerable to issues of access, capabilities, and utilization of ICT than ordinary citizens due to their physical, regional, economic, and social conditions, including the illiterate, elderly, and farmers and fishermen who lack education [13, 14]. The technical difficulties in accessing economic, legal, social, and government support due to the inability to read or write about many means and media provided by ICT society are generally referred to as a phenomenon in which access to knowledge and information differs by economic class, gender, and age [15]. If the information divides result from inconvenience caused by the inability to use ICT, failure to use information can spread to economic, social, and cultural gaps and lead to various inequalities [16]. The lack of information capabilities does not lead to a digital virtuous cycle of information utilization, application, and beneficiary, which promotes deepening inequality (**Table 2**). Therefore, in order to realize the digital inclusion and digital virtuous cycle of the intelligent information society, pan-government policy promotion, social consideration, and agreement are needed to bridge and alleviate the digital divide of the vulnerable (see **Table 2**).

## 4. Public service and digital inclusion in non-face-to-face society

COVID-19 has caused a dynamic society to disappear. Personal hygiene has become of paramount importance and face-to-face private and official meetings have decreased. Media, work, and education suited to the hyper-personalization era take place at home through high-speed, ultra-advanced devices, and most lifestyles revolve around small apps on the display screens of smartphones or computers. In this era of New Normal, untact is causing us to face a turning point in the familiar forms of life [9]. It is time to think about efficient ways to increase non-face-to-face communication between humans and machines with AI. The acceleration of digital transformation is alienating the socially vulnerable, such as the elderly, low-incomers, and disabled, and there is a serious digital divide between the level of informatization and digital capabilities [17]. Therefore, social responsibility is more required in the transition to the 'untact' era. For example, the financial sector is developing a 'finance for the elderly, the disabled, multicultural, and the illiterate' program at commercial banks. We are trying to meet the needs of enthusiastic social contribution activities such as smartphone and mobile banking app usage education, free Wi-Fi installation, reading big letters of our app, 24-hour chat service, etc. that meet rapid changes in the financial environment. This is why many social and public sectors should focus more on ways to bridge the financial digital information gap and strengthen countermeasures, judging that the increase in non-face-to-face services (digital humans, kiosk, etc.) will dampen the activities of the underprivileged [9]. Among the wide assortment of Artificial Intelligence technologies, chatbots have seen a significant uptake by public agencies. Chatbots are deployed across a range of public services ranging from immigration, law enforcement, health, and transportation to utilities and numerous others [4]. They emulate conversations with humans using natural language processing capabilities, enabling them to recognize requests and facilitate text-based or voice-based dialogs [18] to respond to service enquiries and predict user behavior based on previous inquiries of a similar nature. Variations of chatbot interfaces are driven by the user's device of access (e.g., mobile phone, laptop) and the platform (e.g., messaging app, Web page) that supports the chatbot [19].

On January 13, 2022, the Ministry of Science and ICT held a public hearing on the enactment of the Digital Inclusion Act and stressed the need for legislation. The Digital Inclusion Act, which consists of a total of 6 chapters and 35 articles, stipulates the state's responsibility to bridge the digital divide. The main provision is that the government will diagnose the level of digital competency of the entire nation and establish a digital competency center to provide education. Representative Kang Byung-won, who proposed the bill, said that as the non-face-to-face culture spreads due to COVID-19, digital transformation has accelerated, and the purpose of the digital inclusion law is to allow all citizens, including the low-incomed, disabled, and elderly, to enjoy sufficient benefits. Since 2021, the Ministry of Science and ICT has striven to build digital infrastructure by opening digital learning centers and installing public Wi-Fi on city busses and subways. However, despite the government's political efforts, the level of competence of the disabled, the elderly, the low-income, and the farmers, who are defined as vulnerable with respect to information compared to the general public, was low.

According to the 2020 survey of the vulnerable class of information released by the Ministry of Science and ICT last year, the level of capability, an indicator of the basic ability to use computers and mobile devices, was 60.3%. This means the level of the vulnerable class when the level of the general public is 100. Even in the "utilization" index, which measures the degree of quantitative and qualitative utilization of computers, mobile devices, and the Internet, the level of the information-vulnerable class was only 74.8% [20].

## 5. Research method and results

### 5.1 Research method

Five experts on interviewing conducted interviews to learn how to utilize a voice-based chatbot civic service to increase its accessibility to vulnerable groups and citizens, and the attendant challenges and opportunities. Five experts in charge of developing AI-related tasks in realizing interactive interfaces and services were asked to answer by e-mail between August and December 2020 due to COVID-19. The composition of the survey question was extracted by considering the application plans of voice-based public services based on the usage cases of public sectors in the ease of information accessibility of the vulnerable classes (see 2.1). The survey had several categories defined as follows: Technology as the installed voiced-based chatbot, efficiency as the degree to which the chatbot facilitates service delivery while minimizing the cost and resources required [21], effectiveness as the degree to which the chatbot produces an outcome that is supposed to be produced (attaining the goal of using the chatbot) given the resources invested when it is used in delivering a service [22], and Openness & Accessibility. Openness is the degree to which chatbots transparently disclose their identity to users before a service interaction starts and provide the rationale (to users or their representatives such as customer advocacy groups) for decision-making when delivering a service [23]. Accessibility is limited by the term "information accessibility" means ensuring that information and communication services provided through application software installed on websites and mobile communication terminal devices are easily available (Seoul Metropolitan Government Ordinance No. 7232, 2019).

The online survey examined whether citizens were receptive to being provided voice-based services. Interview participants were drawn from Seoul, Gyeonggi, Busan, Daejeon, and Ulsan. Opportunities to promote access to intelligent public administration services, such as requirements and challenges from the perspective of technology, law, or organization, were elicited, and specific ideas for potential applications and voice-based civil services were also requested from the perspective of publicity. The description of the answer is (1) Technical matters include the technical characteristics of each business entity in the answer, which may be unique business confidentiality of each company or may not be common or priority for all business entities. The answers were extremely personal, so they could not be considered representative, and all of them were anonymously considered because they could not be promoted mutually beneficial voice technology by affecting other sectors. (2) The respondent's personal information has limitations in disclosing various personal information, so only general related information is notified. (3) It is notified that only general answers that can be grouped in common are effectively classified and entered, excluding the interviewer's unusual answers or extremely personal thoughts. There were still many limitations in information-, sharing on common topics (hierarchical approval lines, lack of accumulated data, privacy problems, etc.) that were classified based on the responses and statements of interviewees through research analysis methods and clustered technologies, requirements, and applications (**Table 3**).

Q1. Definition: P1, The voice bot is a chatbot in which a virtual secretary has a voice (audio option, interactive interface). In other words, the interaction is much more dynamic and immediate because it can talk like a virtual assistant instead of writing. It should be considered that people are much calmer when speaking than when writing. P2, It is an intelligent assistant using natural language. It is a natural language understanding (NLU)-based voice channel for communication that operates by converting voice into text format. P5, is an interactive interface that uses apps, messaging platforms, social networks, or chat solutions for conversation.

Q2. Technology Advantages: P1, the technology used varies depending on the type of voice bot available, and some include more steps than others depending on what you are trying to do, but the process of the oral-to-written transition is relatively

**1. Classification of experts interviewed**
P1 Software Project Developer
P2 Director of Software Design
P3 Local Government Artificial Intelligence-based Policy Officer
P4 Technology Innovation Center researcher
P5 AI magazine contributer

**2. Common question.**
Q1. as Definition, What's a voice robot?
Q2. as Technology advantages, What technology is used for voice bots?
Q3. as Efficiency, Why do you need voice chatbot service for citizens?
Q4. as Effectiveness, Why should the government introduce voice chatbots in public service?
Q5. as Openness and Accessibility, If the government introduces voice-based services, what points can be useful including AI Technicals?

**Table 3.**
*Interview participants and common questions.*

simple. When the bot receives audio input and converts it into text, AI processes it. For example, after converting audio files into text in real time using Google's Speech-to-Text tool, natural language processing and deep learning techniques are run to read and analyze the stimuli, and then perform knowledge-based processing. When everything is analyzed, the answer is found and sent to the user. P2, Identifying the key markers of the speech and concluding the best response to the conversation or query, the TTS (text voice conversion) engine completes the interaction by converting the response into audio or voice. It is trained to complete the entire speech understanding and response process, which is almost human-like, and a speech assistant is an intelligent communication method. P4, It can be easily integrated without additional channels in partnership with the Google Cloud Contact Center. P5, The interactive secretary automatically tracks scheduled tasks, networks with other chatbots, and improves the ability to process client queries. Voicebots are being developed to provide a wide range of simple answers to basic questions, smooth and beneficial support, and are trying to significantly minimize the time it takes to handle questions.

Q3. Efficiency: P1, P3, P4, and P5. ① In interaction with citizens, it can be provided at a high speed without the need to write or click any more buttons. ② People are not as careful when they speak as when they write. As a result, this carelessness leads to a new request/question method, which can rapidly increase the amount of knowledge-based data, increasing diversity in command analysis, and thereby improving deep learning. ③ Increasing the efficiency of the bot allows public officials to focus on dealing with more complex problems and eventually improves the quality of services they provide to citizens. ④ The use of voice makes it a means of social inclusion, as it is easily accessible to people who are unable to write, are unfamiliar with using technology, or are those with disabilities, the elderly, other cultures, and farmers and fishermen. ⑤ Customer satisfaction can be increased by providing more options to promote interaction with citizens. P2. ① Voice is the most natural communication mode and is easy to use, making it the most preferred communication mode. ② In the latest smart devices, voice bots are a natural alternative to existing devices rather than conventional touch-tone methods. ③ A wide range of customer services can be provided through the knowledge base and back to personalized services. It can also be integrated with numerous smart speakers such as Alexa and Google Home.

Q4. Effectiveness: It provides P1, P2, and P3 citizens with more personalized, intelligent, and insightful interactions in the following major areas. ① Provide a highly personalized service experience that provides timely and appropriate information to citizens. ② Use advanced machine learning models to enable smart matching to connect the best agents to each citizen's interaction. ③ Use strong AI technology to create smart interactions. ④ Integrated, authorized agents are authorized through AI-based guidelines for content and proposed tasks to help expedite public service processes. ⑤ It reduces complexity in simple business operations for both citizens and public institutions. P4. Public institutions can use chatbots to connect with citizens, engage various stakeholders to solve social problems and use a powerful tool called sophisticated chatbots to provide citizens with personalized solutions, receive immediate feedback, and collect information in real time. Real-time citizen response measurements are possible, and the use of chatbots can automate work to enable labor movement and contribute to narrowing the digital information divided.

Q5. Openness and Accessibility: P3, Bots provide consistent results on a regular basis for the provision of general information by public institutions without sleeping or taking a break, communicating with customers, and improving citizens' participation. P2, Digital Twin will provide a single interface for recognition and operational control of jurisdiction in the future. ② Blockchain has the potential to innovate government services by providing transparent records of government transactions and reducing friction between public and private sector ecosystems. It can change many government functions, including identity, voting, public records, and the regulatory supervision of procurement and supply chain risks. ③ Interactive platform functions are needed for data exchange. It creates valuable data streaming for public and individual users and allows data collaboration between governments, industries, and citizens. ④ It uses technologies, including the Internet of Things (IoT) and AI, to provide new ways of working, reserve resources, reserve and coordinate facility services, and provide improved experiences such as information sharing and collaboration.

## 6. Online survey

In all, 112 ordinary citizens(respondents) from 20s to 60s were asked by an online survey agency to conduct a survey in 2020 concerning the premise that "providing AI (Artificial Intelligence) voice-based chatbot services by government agencies is necessary for citizens to communicate with the government." With a total of 10 questions, only citizens' attitudes and intentions toward artificial intelligence voice-based chatbot services were checked with simple questions about Personal (6 questions) and Social(4 questions) Factors. We expected there to be enough sub-questions or variables for each question, but a deeper investigation would not touch on matters deeply related to the thesis that this paper argues, so we will discuss further questions or variables for the experiment later. In terms of technology, acceptance was considered the important variable, only the age group was checked except for the levels of education or gender, (the groups were numbered 0: those in their 20s (2 persons). 1: those in their 30s (8 persons), 2: those in their 40s (34 persons), 3: those in their 50s (39 persons), and 4: those in their 60s (29 persons). A five-point Likert scale ranging from 0 to 4 of *Very not good*, *Not good*, *Normal*, *Good*, and *Very good* was used.

1. In the matter of Personal Usage, the following questions were rated: Q1. I am confident in understanding and using AI voice services in government agencies. Q2. AI speakers are suitable for use by government agencies. Q3. AI speakers are innovative for government agencies. Q4. Using AI voice-based chatbot service

is very interesting and fun. Q5. If I use AI speakers in government agencies, the use of government agencies will be more useful to me. Q6. If I use AI speakers in government agencies, it will be more convenient for me to use them.

Personal level survey results (see above No. 1.): 33.9% of those responding to Q1 said that they can make positive decisions about their capabilities and use them for their own purposes without any difficulty in using the technology. 28.6% said yes, 17% said yes, indicating that the technology's positive effects are much higher. The Q2 for value suitability is 41.1, 28.6, and 16.1%, indicating that the relationship between the recipient and the technology also has a very positive perception. For value innovative, 32.8% of Normal 32.1% Good, and 20.5% Very Good, indicating that they have a very positive attitude toward interest in new products or services. In terms of interest in use Q4, which measures the degree of satisfaction and interest in the behavior they use, were 34.8% Normal, 30.4% Good, 25.9% Very Good, and very little about 1% of the total was Not Good. Therefore, it seems that most of the survey questions are interested in voice AI services. For Q5 and Q6, the results for the responses regarding acceptance showed that the government had a positive perception of and attitude toward the use of AI voice-based chatbot services at a comprehensive personal level, including 37.5% Good and 35.7% Very Goods in terms of personal convenience and usefulness.

2. In the matter of Social Usages, the following questions were asked: Q7. The use of AI speakers in government agencies should be aligned as a social flow and phenomenon. Q8. The use of AI speakers by government agencies reflects the social image of our society and should be used. Q9. If AI speakers are used by government agencies, the use of government agencies will be more socially useful for everyone. Q10. If AI speakers are used by government agencies, the use of government agencies will be more socially convenient for everyone.

Social level survey results(see above No. 2): For social conformity, Q7 is 33.9% Normal for subjective norms, 31.3% for Good, and 20.5% Very Good. Thus, the influence of social sympathy on individuals is beneficial and positive, and for social image, Q8, is 36.6% Normal, 25.9% Good, and 21.4% Very Good, so they seem to have a positive perception of technology acceptance as the degree to which their social status or image within the social system is increased or strengthened The responses to Q9 are 36.6% Good, 29.5% Normal, and 17.9% Very Good. Finally, for Q10, social convenience is 38.3% Good, 27.7% Normal, and 18.1% Very Good, indicating that citizens maintain a very positive perception and attitude that the overall social level of utility and convenience is high.

## 7. Results

The preliminary survey of ordinary citizens found that voice-based chatbot services generally had positive expectations for personal usages and satisfaction of many people, and were considered excellent in terms of their usefulness, utility, convenience, and validity for government use. In social conformity as subjective norms, individuals showed a beneficial and positive attitude to social sympathy, maintaining positive perceptions of technology acceptance reflecting the social image and social status reinforced. As a result, both aspects have similarly reacted positively and constructively to the rest for usefulness, utility, convenience, and validity.

In interviews with the group of experts, P1, P2, and P5 stated that the voice-based chatbot service is an interactive interface that combines text and speech based on natural language understanding to serve as a virtual assistant through speech recognition. P5 stated that the interactive assistant automatically tracks scheduled tasks and improves the ability to handle client queries by networking with other chatbots. Chatbots can move from simply being able to answer basic questions to providing seamless and beneficial support to all types of users, dramatically minimizing the time it takes to solve questions. Therefore, P1 and P2 responded that they could quickly interact with the needs and needs of citizens and that they were helpful to the elderly, the disabled, and the illiterate. As a result, P4 stated that first, institutions are delegating ordinary and everyday tasks to chatbots to reduce employee workload and response time, and second, public institutions are using chatbots to connect with citizens and engage various stakeholders to solve social problems. Third, public institutions can use chatbots to receive immediate feedback, understand citizens' perspectives on problems, and gather information in real time to hear citizens' voices about the problems faced by local communities. Chatbots are also likely to experience similar trends and will continue to learn how to adopt and use these new tools. P8 and P5 describe the field of use of voice service introduction by government agencies, and bots can communicate and interact with customers because they can provide consistent results 24 hours a day. ① The digital twin will provide a single interface for future jurisdiction awareness and operational control. ② Blockchain provides transparent and authoritative records of government transactions, reduces friction between the public and private sector ecosystems, and allows innovations in government services such as regulatory oversight of identity, voting, public records, and procurement and supply chain risks. ③ An interactive platform function for data exchange for commercial and social benefits is possible that creates valuable data streams for public and private users and allows data collaboration between governments, industries, and citizens. ④ Technologies such as the Internet of Things (IoT) and AI can be used to provide new ways of working, reserve resources, schedule and coordinate facility services, share information, and collaborate. ⑤ Applications include industrial and process-specific situations, such as employees who use digital pens that interact directly with backend processing systems beyond productivity gains, or patients monitored remotely through wearable interfaces at home to improve treatment.

As a result, the interviews found that as an intelligent virtual assistant using voice interfaces, chatbots are technically more convenient in saving time and providing verbal services rather than writing sentences. They greatly contribute to enhancing the utility of citizens' interactions and providing social digital inclusion, which helps bridge the information gaps among the socially vulnerable. In terms of effectiveness, the system functions by ensuring that information is integrated and strengthened through machine learning and deep learning to provide personalized services, help make decisions quickly and easily in solving social problems, and maximize the availabilities of labor. It can be integrated with all AI technologies in openness and accessibility to provide new services and provides stronger technology to prevent crimes or protect personal information through regulation, perception, or prediction, and affording means to protect the socially vulnerable.

## 8. Considerations of government agencies with voiced chatbot service trials of voiced chatbots for public services in Korea

With the proliferation of powerful hardware and soft-computing techniques, an era of AI-based services has emerged in recent years. AI services have witnessed a new industrial revolution across the globe. Today, AI is becoming part of real life. With the evolution of technologies of deep learning and big data, decision-making, data analysis, and smart assistance services that integrate information from different divisions have also come into use in government and the public sector. Chatbot using artificial intelligence may ease human life by knowing real-time news or collecting information, suggestions, and recommendations, and making reservations and preparation in advance (see **Figure 1**). The voice recognition technology of AI chatbots in cloud computing is rapidly evolving through deep learning, which functions to strengthen access between citizens and government, thus promoting government transparency and preventing corruption, which is recognized as essential functions for the role [9]. Since November 2021, the government has been expecting to create new values and improve efficiency based on public intelligent information systems by Corp. Sorizaba providing voice recognition AI services that apply voice synthesis technology to AI voice recognition solutions of integrated management systems [24].

Here we discuss some trials in Korean governments and public sectors (see **Figure 2**). In 2020, the Ministry of Public Administration and Security announced that AI can automatically complete difficult and complicated forms of complaints and provide direction regarding the public services citizens want when they simply speak. In particular, Jeju Island will provide these services by selecting about 30 types of documents issued with high frequencies to the socially vulnerable first, particularly those who have difficulty with such documents. Daejeon City and Gyeonggi Province

**Figure 1.**
*Development of the digital ecosystem supporting the government system. Source: Song [9].*

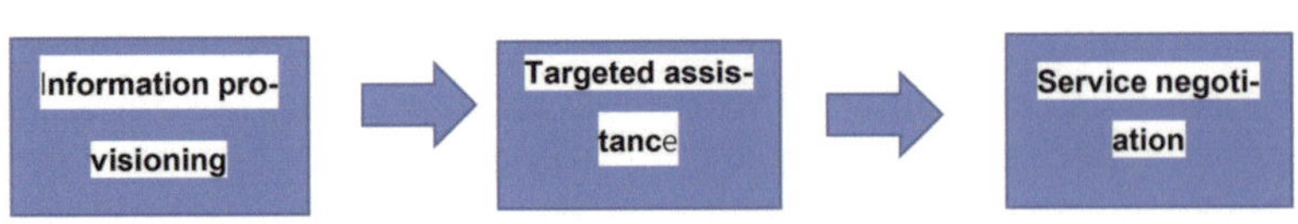


**Figure 2.**
*Chatbot-mediated public service delivery levels. Source: [19].*

will introduce "Smart Mirror," a safety inspection service using a combination of smart glasses and IoT technologies, and an AI that allows citizens with disabilities to receive various forms of information quickly and accurately. By learning sign language movements, information can be provided in sign language for the hearing impaired, and voice guidance for the visually impaired is also possible [9]. Pol-Bot technology is being developed for application in 182 police complaint call centers (182 call centers): 112 is for calls that require emergency police dispatch, while 182 is for police complaints and advice not related to crimes, while Polbot handles traffic fines, fines, unmanned crackdowns, penalties, suspension, traffic accident investigations, misdemeanor crime amounts, and missing reports. It has developed a voice-conversation-based chatbot dedicated to 182 call centers with ETRI, the National Police Agency, and domestic AI companies [9]. Similarly, the "Civil Complaint Government 365" service in Korea is an interactive civil complaint counseling service provided by six administrative agencies (National Police Agency, Korea Customs Service, Military Manpower Administration, Forest Service, Ministry of Unification, Ministry of Public Officials Pension Service). The interactive civil service (National Secretary "Gubbi") provided by public institutions (Public Officials Pension Service) opened in May of this year and provided chatbot services to KT GiGA Genie Internet subscribers as an AI speaker in 2021 through a cloud-based system [25]. Collaboration between humans and AI is particularly emphasized in the decision-making area. In an environment where the pace of change and uncertainty is increasing, the frequency and complexity of decisions made by companies and individuals will increase, so the need for artificial intelligence in decision-making will increase [26].

## 9. Considerations when designing voice-based services

In order to improve the quality and convenience of administrative services, institutions are emerging that provide counseling services after working hours using voice-based chatbot services in real-time. Despite the remarkable development of AI technology, concerns remain over the possibility of service levels falling short of expectations or making wrong decisions due to misunderstandings in natural language and data processing in each sector. In terms of technology, there are not only difficulties in ontology generation, knowledge data construction, and big data, but also in non-technical perspectives that lack a venue or opportunity for discussion and academic knowledge exchange [2].

If so, it is necessary to find out all the considerations to be taken into account for the socially vulnerable to realize the provision of accessible interactive public services. First, the human-computer interface method has developed greatly since computing technology was introduced. Its foundation evolved from text to graphics, and a graphic-based interface (GUI) was developed that provided a form of mouse clicking or finger touching a specific application icon, followed by a voice recognition-based interface (VUI; [27]). Therefore, speech recognition technology converts a person's voice into text data form suitable for purposes and target classes. In order to increase the accuracy of speech recognition in the voice input process, a pre-processing technology that filters out noise and unnecessary data from data collected by a recording device is also used. Voice interfaces using technology should be designed to meet the following requirements (see **Figure 3**).

First, short and accurate answers from each voice assistant are important to easily achieve the user's purpose of use, taking into account the cognitive limitations of

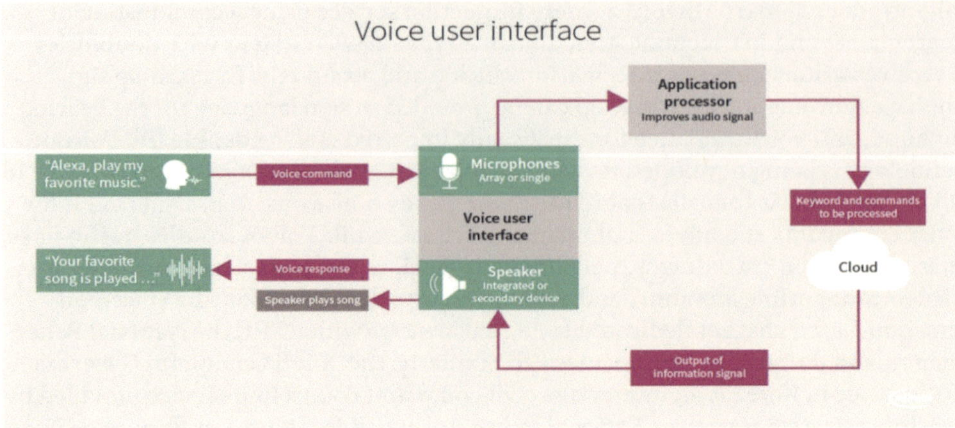


**Figure 3.**
*Voice recognition interface-based service configuration. Source: https://www.infineon.com/cms/en/applications/consumer/smart-speaker/*

citizens' education level, disability, and age level. When the conversation is long or there are more than three requirements in the process of answering, specific fatigue with the voice assistant may stand out [28].

Second, speech recognition differs from past interface methods (where specific instructions or icons were connected to specific apps) in several respects. Voice recognition-based devices should be able to understand user commands through natural language processes and recognition first. In other words, it should be possible to accurately convert the recognized voice into a command while the sound can be recognized as human speech. Based on the recognized command, it is then necessary to invoke a corresponding function or service connection operation [27]. Voice interfaces are largely made in the order of language understanding, conversation processing, and response generation (see **Figure 3**). Language understanding refers to the process of analyzing a user's speech and converting it into a machine-intelligible semantic structure, and creating a semantic structure of a system response by applying an appropriate dialog model is called a dialog process [29]. This is a priority to be addressed in establishing knowledge-based services such as process integration, information exchange and sharing, and the limitations of the laws and regulations of each ministry.

Third, the interconnection with other information systems is crucial [30]. For example, the mobile payments system is changing by combining POS (point of sale information management system) solutions with business back office platforms, and establishing an application programming interface (API)-based personal information analysis and management ecosystem. Trends and changes like these are in response to the observation of falling profitability in traditional payment methods, and payment companies in advanced countries are sharing customer information through open banking and API-based solutions and making efforts to generate profits using them. The back office system is a procedure that consumers and suppliers must recognize as a process of co-creating value, and it is possible only when a comprehensive legal framework and strategic cooperation of each platform operator are guaranteed.

Fourth, the integration with voice support services (Amazon Alexa, Google Assistant, Microsoft Cortna, Apple Siri, etc.) used in users' daily lives can be needed to enable. The value chain of the voice AI market allows it to lead to the base technology, voice AI platform, hardware with voice recognition functions, and services available to users [29]. For example, during breakfast, we often receive the weather forecast for the day through voice support services. In order to increase the convenience of public NURIJIB, which means a website in original Korean, that are frequently and irregularly visited, basic services such as mortgage loans, pensions, guarantee requirements, or inquiries will be installed so that they can be used without a separate custom extension.

Fifth, strengthening accessibility to the social vulnerable, which is weakness revealed by e-government, is an important requirement in implementing digital civic services. A public institution emphasizes the functional role of the corporation in ensuring equal social opportunities through effective and practical policies and financial support, especially in terms of social consideration and social distribution for the vulnerable. Therefore, it is predicted that it will provide an environment that naturally controls digital devices with human senses, actions, and cognitive abilities so that people with severe disabilities and severely ill patients can be guaranteed the right to access data and protect personal information. With the development of artificial intelligence sensor technology, the interface that is attracting attention is NUI(Natural User Interface)[2], as it allows users' actions to be directly input through communication(like human) methods. It is developing into an AI, multi-touch, and largely augmented reality (AR; [31]).

Sixth, as a personal information processing problem, handling and protecting information-related privacy is an important topic in the AI era. The need for the user's authentication is natural for several advanced digital government services that include information whose confidentiality is crucial to privacy protection. The two-step authentication technique after voice recognition by combining a cryptographic method containing a personal pin and an individual's unique voice characteristics displays each prompt that verifies an authentication request with a user authentication app if authentication is required using the user's personal device. In addition, if "Biotech + Certificate (PKI)" is combined and used as a speaker authentication technology, authentication procedures based on voice authenticity [32] can be implemented. In the case of public institution users, they can verify their identity using their personal taxpayer number. This is sufficient to be used in various governmental services because the possibility of abuse is very limited [33].

Seventh, this task involves understanding the standard language of speech recognition. For example, to receive directions provided by the app, one can say the starting point and destination in a local dialect. Since the vocabulary is limited, this raises concerns about whether even local dialects can be understood. The voice conversation interface (**Figure 3**) of the voice recognition-based service recognizes voice with a customized voice recognition engine. P1 believes that although understanding local dialects is an important requirement for accessibility and acceptance of various classes and local citizens, it will have little relevance to the financial, private, and government services below in that they are expected to use the standard language. In addition, considering that about 70% of local dialect speakers can accurately

[2] In computing, a natural user interface, or NUI, or natural interface is a user interface that is effectively invisible, and remains invisible as the user continuously learns increasingly complex interactions. The word "natural" is used because most computer interfaces use artificial control devices whose operation has to be learned. Examples includevoice assistants, such as Alexa and Siri, touch and multitouch interactions on today's mobile phones and tablets, but also touch interfaces invisibly integrated into the textiles furnitures (Wikipedia).

recognize standard language and that financial and administrative languages are the standard language, they may have difficulty with some special jargon, but this is not recognized as a problem.

Eighth, there is a need for API (now open API is operating) to link and integrate the data of government ministries and public institutions and to access information. It is imperative to realize more sophisticated interactive services and to resolve problems of information distribution and the lack of linkage in procedural access to information systems related to public data through third-party applications. In fact, various public data and related information systems have been established and operated by government ministries, individual departments, and public institutions, but it is often difficult to utilize such information. Standardization of information systems and data is necessary in terms of data linkage, due to the disclosure of such problems as not being able to use user information in the past during business processing if information is disclosed and disappears due to the fragmentation of information systems.

## 10. Policy suggestions to mediate voice-based services for digital inclusion

The artificial intelligence-based conversational chatbot platform continues to develop, designed on purposes and expected to interact with the public in multiple languages, analyze sentiments and intent, and collect and analyze data to provide users with a personalized experience. The platform should be able to provide assistance with general tasks, including but not limited to making use of government services and gaining access to functions such as registration, login, reset the password, department service relation information, event information, (and) new launches like requests for proposal documents. The chatbot should be able to turn speech into textual input and turn text into speech. Platforms should also have other deep learning capabilities, which can be used by platforms for the discovery and recommendation of eligible schemes and services to the users of bots/applications. Users should be recommended the best-suited schemes and services available in the government domain based on their profile as well.

### 10.1 Provisions for customized public services of users

First of all, by combining blockchain technology and voice-based services, official documents are requested using vast amounts of public data [34]. In terms of the structure and security of distributing and storing public data disclosed in the blockchain, all blockchain participants can share and own information. For example, citizens are asked for basic personal documents when they want to apply for a lease loan. Official documents such as employment certificates, family certificates, lease agreements, and health insurance fact-finding certificates must be submitted to other institutions or companies from time to time. Currently, the digital government is strengthening user authentication and innovating public and private services with safe and convenient digital identification online and offline through mobile identification cards with a system that recognizes personal information through the voice [33].

Second, it provides information search and counseling on related loan screening methods, laws or taxes, rates, bond transfer, and guarantee insurance [35]. Robot automation systems and voice-based chatbot services provide information including daily laws or local housing finance corporations for citizens, such as those related to taxes and specific problems or policies. If citizens have to access the data storage after voice recognition to determine whether or not they are eligible for certain services such as health insurance reduction or housing subscriptions, then the expected cost of related services can be provided to users through voice output. Most civil service services had different patterns of processes. It is expected that the pattern of complaint handling can be divided into several groups via machine learning classification techniques using various data generated in the complaint handling process. If this is systematized, civil petition processing can be automated according to the classification according to characteristics [13, 14].

Third, it uses big data and the Internet of Things (IoT) to notify citizens' of information collected through algorithms or provide personalized services for reminders [35]. In the future, the government plans to introduce a personalized "national secretary" system that supports applications for processing. For example, "OOO is eligible for principal repayment this year," "Your child OOO is eligible for a national scholarship application and can apply until x month x month," and "You can apply for a home loan." It will be serviced through a private messenger app or AI speaker by inserting basic functions into the "Government 24" and opening them with an open API.

Fourth, the government can use MyData to search, store, and distribute personal information held by public institutions to facilitate the convenience and implementation of citizens' data sovereignty. For example, when applying for policy funds for small business owners, 20 types of required documents can be submitted as MyData service for use for the credit management and asset management of citizens, or loaned to financial institutions as MyData [33].

### 10.2 Enhancing personal security for the vulnerable

Just like any new technology, chatbots come with cyber-security risks. For this reason, it is very important to know the security practices that are commonly implemented when working with chatbots. For the most part, chatbots do not present security issues that have not already been discovered and properly mitigated. The era of big data enables information collection, processing, storage, and utilization, spurring efforts to secure regulations on "global data security" as the decoupling/coupling phenomenon of the customer value chain is accelerating in a non-face-to-face society [36]. That phenomenon requires regulations on the unauthorized use, occupancy, and utilization of stored domestic and foreign big data. In particular, there is considerable concern that it will be abused by exposure to phishing crimes (including smishing, voice phishing, and farming). Therefore, through intelligent monitoring systems and location information systems using artificial intelligence technology, the preventive role can be increased by monitoring, recognizing, and analyzing the circumstances of the socially vulnerable to protect them from such crimes or risks, as all work processes are stored in digital form in a platform-based cloud and audit traces are left [37]. There are also Data Trusts, in which the trustee manages the data and data rights of individuals and groups. In order to introduce the concept of data trust, the role of data trust agencies is important in setting processable information by different stakeholders and using data safely without excessively violating the rights of data subjects. Trusted by the settlor. Authorization must be based on accurate processes, the number of consignors, and data through various incentives and performance sharing that can facilitate data consumers' engagement rather than simply "protecting" data [38]. In other words, socially vulnerable groups are easily targeted for crimes in which personal information is stolen and exposed to crimes, property losses, and physical threats, and the difficulties they face in financial transactions, public administration, and welfare services are increasing. Therefore, preventive measures can be taken by linking the police and financial authorities with intelligent monitoring systems by recognizing and analyzing voice instability in identification such as monitoring and voice biometric authentication based on the database of big data in financial transactions. In addition, it is particularly necessary to share and cooperate with financial institutions by strengthening security policies to compensate for weaknesses of vulnerable groups in the non-face-to-face era.

## 10.3 Policy supports

Citizens' use of AI requires trust and social consensus by establishing a legal safety net in terms of data processing problems, exposure and utilization of personal information, and social resistance and conflicts due to the replacement of manpower by AI. I thus offer suggestions on this point.

First, it is necessary to reorganize and enact appropriate laws and systems. The Three Data Laws, including the Personal Information Protection Act, the Information and Communication Act, and the Credit Information Protection Act (2020.8.5), as well as the Framework Act on Intelligent Information Society and the Electronic Government Act (tentative name; [39]) for enhancing public communication from AI chatbot's functions must be enacted. In addition, the development of AI ethics and guidelines is more urgent than anything else. Responsible AI must be implemented publicly through the adoption of Ethical Standards of AI [9]. In addition, it is necessary to promote and strengthen the self-regulation of large platform companies.

Second is the establishment of a governance infrastructure of the (public-private partnership type based on a cloud platform. It is necessary to quickly introduce a cloud system that can utilize AI technology in all public institutions, including the government. This can increase productivity, efficiency, and accountability and reduce costs.

Third, private companies should identify their needs and prepare tax benefits and funding and incentive systems to foster and promote big data businesses [40].

Fourth, to expand the scope of research as a basic academic and strengthen AI education [41], democratic citizen digital literacy education and data expert training are comprehensively needed. The power to respond to the many side effects that will occur due to the introduction and expansion of artificial intelligence should be strengthened to make AI artificial intelligence for all citizens. Critical thinking and cultivating social responsibility can cultivate ethical judgment or self-aware communication skills and the evaluation of text that promotes and incites dangerous information, biased or ideological thinking, and conflicts.

## 10.4 Implications of overcoming the digital divide with voice-based public services

First of all, the introduction of voice-based service chatbots to support the socially vulnerable will have a profound effect on the change in the structure of the civil service organization and decision-making structure. The vertical command system (top-down) makes it difficult to perform and inefficient for work delivery. Intelligent government models formed based on data require cooperation and the exchange of information from each department, and integrated information services cannot be provided by decentralized departments. Therefore, organizational change will contribute more to our society than anything else by providing an efficient voice-based service with a rich database, simplified procedures, and horizontal and neutral organizational decision-making.

Second, from the standpoint of digital inclusion, information accessibility means guaranteeing the human rights and basic rights of the socially vulnerable. The Open Data Strategy (ODS), which is being promoted mainly in the EU, holds that such institutions as big data and open government data should be considered to have started by expanding information access rights as basic rights rather than as serving to activate big data or markets. It should be seen that Korea also started with the expansion of the right to access information as a basic right to broadly guarantee the right to use public data. Korea has also implemented the Act on the Provision and Promotion of Public Data, and the right to access information to Open APIs means "the right to freedom of information or the basic right to know," similar to the European Parliament's "recommendation on access to information and freedom of information" or the 2013 EU [42].

Third, constructing voiced-based chatbots serves to ease the knowledge and information gap of the socially vulnerable and to guarantee the human rights of the disabled. The development of digital technology, which spurred the advent of an information society, has created a new communication environment that provides more information to many people and spreads various cultural experiences. However, for the socially vulnerable with physical disabilities or economic difficulties, these environmental changes also have a negative aspect that they can act as barriers to normal social participation rather than an opportunity to improve their quality of life [43]. For example, as the function of broadcasting media expands, concerns are raised that broadcasting media may intensify the isolation and alienation of the disabled who are restricted from using such media [44]. Behind the progress of the information society, the gap between the rich and the poor has widened depending on the production, ownership, and accessibility of knowledge information, widening social polarization and resulting in mass production of the underprivileged [43]. This term is most commonly used to refer to the distribution or utilization of information inequality in the information divided and the digitally divided society. The "digital divide" means the gap between opportunities to use information and communication in a digital environment and the ability to use it [44]. At the center of the problem are the disabled, especially those among the socially vulnerable. It is highly likely that the disabled will be excluded from various forms of social participation as they become alienated from means of communication or information acquisition. Social alienation arising from the lack of means and methods such as channels provided by digital media or devices is a problem that society must overcome in the future.

Fourth, it is necessary to prepare measures for personal information and cybersecurity and strengthened digital privacy policies. The greatest vulnerability of AI technologies is hacking. There are various authentication methods such as passwords, PINs, smartphones, biometric authentication, text authentication, and QR authentication, but a "Digital Identity" service authentication method that requires user identification and designation as a consistent government policy is needed [32].

| |
|---|
| 1. Legality: The basis for processing, such as the collection, use, and provision of personal information, must be legal and clear. |
| 2. Humanity: Securely process and manage personal information. |
| 3. Transparency: Personal information processing details are disclosed to make it easier for the data subject to understand. |
| 4. Participation: It has a communication system for personal information processing and guarantees the rights of information subjects are guaranteed. |
| 5. Responsibility: Clarify the management's responsibility for processing personal information. |
| 6. Fairness: Minimize the occurrence of social discrimination, bias, etc. by processing personal information according to the purpose of collection. |
| *Source: The Personal Information Commission [45].* |

**Table 4.**
*Six principles of AI-related personal information protection.*

The Public Key Infrastructure (PKI), a technology that encrypts and signs biometric information such as fingerprints so that they may be safely used by registered users, was set as ITU's international standard in 1989 and is in use worldwide (p. 185). It is necessary to design a combination of speaker authentication and PKI technology to ensure protection because the disabled are exposed to crimes where people who illegally use their information for their own profit, and measures for multicultural families are expected to ease linguistic barriers, the greatest restriction on government use. In addition, autonomous efforts to make regular inspections mandatory are required. The Personal Information Commission [45] announced six principles of AI-related personal information protection and stated that it is mandatory to prepare and implement response procedures for information subject notification, leakage reports, and damage relief support in the process of AI service operation (**Table 4**).

Lastly, the establishment of digital governance is creating a new means for various stakeholder groups to participate at a low cost through the development of intelligent information technology and SNS [37]. Therefore, with the progress of digital informatization, information human rights are being examined [46]. In particular, guaranteeing the right to access information is essential as part of the digital inclusion policy for the socially underprivileged. Being alienated from information directly affects job choice and narrows the scope of economic opportunities, leading to economic poverty. Today, the use of information and information technology is the most important basic living factor and a medium for enjoying all legal interests and basic rights of individuals, meaning that bridging the information gap secures prerequisites for realizing basic rights [47]. Various digital literacy education should also be conducted in various ways to promote the inclusive use of digital governance and digital technology for the socially vulnerable, and it seems urgent to prepare corresponding laws and systems.

## 11. Conclusion

This paper examined the following on the premise that public institutions, which aim to promote citizens' welfare and contribute to minimizing the digital divide for the vulnerable, provide voice-based public services to citizens. First, it was confirmed through a small random online survey that citizens' attitudes and expectations for public institutions ahead of voice-based services were maintained, and interviews with experts with knowledge of AI elicited challenges facing public institutions and governments in strengthening information accessibility. Second, all matters to be considered when establishing voice-based chatbots were investigated in connection with measures to be applied and utilized by public sectors, and third, the implications of voice-based services in terms of public communication and open government services were examined. This is significant in that it provides an opportunity and a foothold for a wider range of citizens to realize their participation in intelligent governments and to practice social consideration and inclusion through strengthening and guaranteeing human rights, basic rights, and information accessibility for socially vulnerable groups.

The interviews confirmed that as an intelligent virtual assistant using voice interfaces, chatbots are technically more convenient to save time and provide services by providing verbal services than written methods, which enhances citizens' interaction in utility and affords social digital inclusion, which helps bridge the information gaps among the socially vulnerable. In terms of effectiveness, information is integrated and strengthened through machine learning and deep learning to provide personalized services, help make decisions quickly and easily in solving social problems, and facilitate the use of labor. It can integrate with all AI technologies in openness and accessibility to provide new services and provides stronger technology to prevent crime or protect personal information through regulation, perception, or prediction, as well as other means to protect the socially vulnerable.

This paper discusses how voice-based public services using natural language can constitute a public communication channel for realizing open government by strengthening human rights and basic rights, digital inclusion, and the protection of information access by bridging the information gap of the socially vulnerable. The government plans to inform the public services in areas close to civic life, including the effective response to the post-COVID-19 era, and ensure social, economic, geographic, and psychological equity in the digital government. In particular, inducing participation of the socially vulnerable is a key task of intelligent government that should be realized. Providing such communication for immigrants, the elderly, low-incomers, and rural people who lack education in natural language using a voice interface helps solve complex problems in the civil service easily. Various considerations in preparing a voice-based public service platform are the prerequisites such as the mutual interconnections of the sectors, NLU, API, and value chains. Policy suggestions are listed for provisions for customized services of users to enhance the personal security of the vulnerable. First, the laws and systems should be reorganized and appropriate measures enacted. Second, the establishment of a governance infrastructure based on a cloud platform is necessary. Third, private companies should identify their needs and prepare tax benefits and funding and incentive systems to foster and promote big data businesses. Fourth, the scope of research as a basic academic pursuit should be expanded and education on AI should be strengthened, including digital literacy. Lastly, the implications of voice-based public services for public communication and open government for the socially vulnerable on information accessibility for digital inclusion against the digital divide, and their ability to further basic human rights were adduced. This research had a major limitation in that it was difficult to find information on the technical and functional aspects of voice-based services, such as academic literature on the integration of the public sector and AI, or research reports from government and related agencies. It was even more difficult to obtain data due to insufficient data management and accumulation, and it is believed that more developmental studies will emerge only when active guidelines for the disclosure and utilization of public data are presented.

In the future, I would like to continue to research social and humanistic influences such as cloud nationality and digital citizenship in line with the 4th industrial revolution, and further consider how to deliver policies effectively and develop channels. The development of comprehensive public communication channels can lead to more transparent government organizations by streamlining decision-making, overcoming the harmful effects of bureaucratic organizations with risks of wrong decision-making, dismantling rigid organizational culture, and reducing the burden on civil servants. In addition, the information technology society using digital technology should never neglect the maintenance of laws and systems for personal privacy and cybersecurity, which should be built and designed around citizens who are not biased toward the government and public institutions in data storage, utilization, and management. AI and the "untacted" governmental system impact our lives directly and provide various standards for assessing the quality of life of citizens dreaming and aiming for a bright and sound future. Efforts on the part of all of us to minimize the marginalized class in the process of implementing intelligent governments are very natural, and we hope that higher administrative services and a stepping stone for a positive paradigm shift can be provided.

## Author details

Jinsoon Song
Dong-a University, South Korea

*Address all correspondence to: songjs@dau.ac.kr

## References

[1] Wellman B. Physical place and cyberplace: The rise of personalized networking. International Journal of Urban and Regional Research. 2001;**25**(2): 227-252. DOI: 10.1111/1468-2427.00309

[2] Song J. Measures and policy suggestions for strengthening the functions of artificial intelligence chatbots to promote public communication. National Assembly Research Service. 2021;**13**(28):339-366 (In Korean)

[3] Business Insider. The ATO's virtual assistant has already answered almost a million enquiries this year. Business Insider. 2016;**12**

[4] Mehr H. Artificial Intelligence for Citizen Services and Government Artificial Intelligence for Citizen Services. 2017. Retrieved from: http://ash.harvard.edu /files/ash/files/artificial_intelligence_for_citizen_services.pdf

[5] William E, Schatsky D, Viechnicki P. AI-augmented Government Using Cognitive Technologies to Redesign Public Sector Work. UK: Deloitte University Press; 2017. pp. 230-248. Retrieved from: https://www2.deloitte.com/content/dam/insights/us/articles/3832_AI-augmented-government/DUP_AI-augmented-government.pdf

[6] Tuzovic S, Paluch S. Conversational commerce: A new era for service business development. In: Bruhn M, Hadwich K, editors. Service Business Development: Strategien - Innovationen – Geschaftsmodelle. Vol. 1. Wiesbaden: Springer Fachmedien; 2018. pp. 81-100. DOI: 10.1007/978-3-658-22426-44

[7] Canalys. Amazon smart speaker ships crossed 10 million mark in Q3(2019). 2019. Retrieved from: https://www.canalys.com/newsroom/worldwide-smartspeaker-Q3-2019

[8] Wirtz BW, Birkeyer S. Open government: Origin, development & conceptual personal administration. International Journal of Public Administration. 2015;**38**(5):381-396. DOI: 10.1080/01900692.2014.942735

[9] Song J. A study on the improvement of public communication and reliability through the introduction of artificial intelligence chatbot by local police. The Journal of Police Science. 2022;**22**(68):239-270 (In Korean)

[10] VSOFT Consulting. 12 global government agencies that use chatbot. 2018. Retrieved from: https://blog.vsoftconsulting.com/blog/15-governments-agencies-that-use-chatbots

[11] Kim H. A study on legal reform measures to protect the personalities of the socially vulnerable. Korea Research Institute of Legislation. 2014 (In Korean)

[12] Kim S. Cultural policy of multicultural society. In: Collection of Academic Conferences of the Korean Governance Society. Vol. 59, No. 6. Korean Governance Society; 2006. pp. 461-470. Available from: https://www.jstor.org/stable/26907117 (In Korean)

[13] Lee C. A study on the quality measurement of public service using process mining: Focused on construction license civil service in N. City Journal of Intelligence and Information Systems. 2019a;**25**(4):35-52 (In Korean)

[14] Lee K. Digital information gap and tasks in the intelligent information

Society. Health and Welfare Forum. 2019b;**274**:16-28 (In Korean)

[15] Min Y. Internet use and information gap. Media Information Research. 2011;**48**(1):158-187 (In Korean)

[16] Joo Y. Intelligent information society and information inequality. KISO Journal. 2018;**33**:33-37 (In Korean)

[17] Bae Y. Shadows in the non-face-to-face era, digital alienation, issues and diagnosis (488). In: Gyeonggi Research Institute. Gyeonggido; 2021. Available from: https://www-dbpia-co-kr.libproxy.donga.ac.kr/pdf/pdfView.do?nodeId=NODE10541164 (In Korean)

[18] Shawar A, Atwell E. Chatbots: Are they really useful? Journal for Language Technology and Computational. Linguistics. 2007;**22**(1):29-49. Retrieved from: https://jlcl.org/content/2-allissues/21-Heft1-2007/Bayan_Abu-Shawar_and_Er icAtwell.pdf

[19] Makasi T, Nili A, Desouza K, Tate M. Chatbot-mediated public service delivery: A public service value-based framework. First Monday. 2020;**25**(12). Retrieved from: https://firstmonday.org/ojs/index.php/fm/article/download/10598/10000

[20] Digitaldaily. Digital alienation is our job... The government is actively seeking to bridge the digital divide. 13 January 2022. Retrieved from: http://m.ddaily.co.kr/m/m_article/?no=229278. (In Korean)

[21] Meynhardt T. Public value inside: What is public value creation? International Journal of Public Administration. 2009;**32**(3-4):192-219. DOI: 10.1080/01900690902732632

[22] Andrews L. Public administration, public leadership and the construction of public value in the age of the algorithm and "big data." Public Administration. 2018;**97**(2):296-310. DOI: 10.1111/padm.12534

[23] Andersen LB, Jørgensen TB, Kjeldsen AM, Pedersen LH, Vrangbæk K. Public value dimensions: Developing and testing a multi-dimensional classification. International Journal of Public Administration. 2012;**35**(11):715-728. DOI: 10.1080/01900692.2012.670843

[24] ITBiz. Establishment of an intelligent AI voice recognition solution for Sorizaba e-government cloud platform. 24 August 2020;**65**:126. Retrieved from: http://www.it-b.co.kr

[25] E-Daily. Saltlux and the Ministry of the Interior and Safety Won the Pan-Government Civil Complaint Consultation 365 Construction Project. Seoul; 29 July 2020. Retrieved from: https://www.edaily.co.kr/news/read?newsId=03388246625839440&mediaCodeNo=257. (In Korean)

[26] Lee C, Seo B, Kwon Y. A study on the effect of artificial intelligence on decision-making: Focusing on the collaboration between humans and artificial intelligence and the personality characteristics of decision makers. Journal of Intelligence and Information Systems. 2021;**27**(3):231-252 (In Korean)

[27] Kim K. A Next-Generation Interface Technology that Connects Impossible to Possible. Vol. 65. Samsung Display Newsroom; 28th April 2020. Available from: https://news.samsungdisplay.com/22952?type=search&category=&search_type=keyword&search=%EC%9D%B8%ED%84%B0%ED%8E%98%EC%9D%B4%EC%8A%A4&search_term=0&page=2

[28] Korea Creative Content Agency. Trends and prospects of speech recognition technology, in-depth reports

of cultural technology. CT In-Depth Report, No. 11. 2011. Retrieved from: https://www.kocca.kr/kocca/bbs/view/B0000144/1756144.do?searchCnd=&searchWrd=&cateTp1=&cateTp2=&useAt=&menuNo=204962&categorys=0&subcate=0&cateCode=&type=&instNo=0&questionTp=&uf_Setting=&recovery=&option1=&option2=&year=&categoryCOM062=&categoryCOM063=&categoryCOM208=&categoryInst=&morePage=&delCode=&qtp=&pageIndex=7. (In Korean)

[29] Samjeong Economic Research Institute. Trends and Business Opportunities in the Voice AI Market. Issue Monitor, No. 126. Seoul: KPMP; 2020. Retrieved from: https://kpmg.com/kr/ko/home/insights/2020/04/issue-monitor-126.html

[30] David S, Craig M, Kaushik L. The way to a robotic process automation cognitive enterprise. New York: Deloitte University Press; 2016

[31] Kim K. A next-generation interface technology that connects impossible to possible. Samsung Display Newsroom. 2020. Retrieved from: https://home.kpmg/kr/ko/home/insights/2020/04/issue-monitor-126.html (In Korean)

[32] Choi U. Cybersecurity of the intelligent government. In: Jung C et al., editors. The Intelligent Government Policy Issues, 125-140. Seoul: Yoon Sung-sa; 2021 (In Korean)

[33] Joint Report of Related Ministries. Korean Version of the New Deal Comprehensive Plan: A Major Transformation of Korea into a Leading Country. 14 July 2020. Retrieved from: https://www.korea.kr/briefing/pressReleaseView.do?newsId=156461589

[34] Delloitte. A medium for promoting the data economy, a public-led open data platform. Industry Focus, 74. 2019. Retrieved from: https://www2.deloitte.com/content/dam/Deloitte/kr/Documents/public-sector/2019/kr_gnps_issue-highlights_20191125.pdf (In Korean)

[35] Information and Communication Technology Promotion Center(ICTPC). Status of AI use at home and abroad, public application. ICT SPOT ISSUE. 2018;**2**:21-30 (In Korean)

[36] FinancialNews. Security regulations and U.S. economic exchanges may be cut off during the "realization of decoupling". 8 September 2020. Retrieved from: https://www.fnnews.com/news/202009081509524917 (In Korean)

[37] Jeong C. Cybersecurity of the Intelligent Government: Intelligent Government Policy Issues. Seoul: Yoon Seongsa; 2021 (In Korean)

[38] Jang K. A study on the introduction of a data trust system to expand the right to self-determination of personal information. Journal of Intelligence and Information Systems. 2022;**28**(1):29-43 (In Korean)

[39] Jeong S, Lee J, Kang J. Exploratory research on the use of artificial intelligence technology in the administrative field. KIPA Research Report. Korea Institute of Public Administration; 2019. pp. 85-100

[40] National Information Society Agency (NIA). Data-driven, establishment of scientific policy tasks, a working handbook. 2017. Retrieved from: file:///C:/Users/sec/Downloads/(170728)_%EC%A0%95%EC%B1%85%EA%B3%BC%EC%A0%9C%ED%95%B8%EB%93%9C%EB%B6%81_%EB%B3%B8%EB%AC%B8(%EC%97%85%EB%A1%9C%EB%93%9C%EC%9A%A9).pdf (In Korean)

[41] Science and Technology Policy Institute (STEPI). 2018 STEPI Annual

Report. 2018. Retrieved from: file:///C:/Users/sec/Downloads/STEPI_2018.pdf (In Korean)

[42] Ahn S. The background and significance of the publication of the Information Human Rights Report. Data Collection of the Joint Academic Conference to Commemorate the Publication of the Information Human Rights Report, 3-13. 2013 (In Korean)

[43] Lee S. Changes and practical tasks in the social welfare paradigm of the information society. Nambu University Paper Collection. 2008;**8**:19-40 (In Korean)

[44] Song J. A study on policy measures for the expansion of broadcasting access rights for the disabled. Broadcasting Research. 2003;**57**:147-178 (In Korean)

[45] Personal Information Commission. Artificial intelligence (AI) personal information protection self-checklist. 31 May 2021. Retrieved from: https://www.pipc.go.kr/np/cop/bbs/selectBoardArticle.do?bbsId=BS217&mCode=D010030000&nttId=7347 (In Korean)

[46] Hwang J, Lee M. Legal reform direction for ensuring information access without discrimination. KISDI Issue Report. 2004;**4**(30):24-26 (In Korean)

[47] Hong S. A constitutional review on the closing of the digital gap - based on the need and method of closing the digital gap from the perspective of the social state principles. Land Construction Method Study. 2009;**44**:475-498 (In Korean)

Chapter 5

# Personality for Virtual Assistants: A Self-Presentation Approach

*Jeff Stanley*


## Abstract

Self-presentation is a sociological line of research relating concrete human behaviors, perceived personality traits, and social interaction goals. According to self-presentation, people engineer their own attributes such as behavior, clothing, and speech to try to affect how they are perceived and to accomplish social goals. Recent studies demonstrate that principles from self-presentation apply not only to how humans interact with other humans but also to how humans interact with machines. Therefore, the rich body of self-presentation research can inform virtual assistant personality and behavior. That is, if a virtual assistant is trying to accomplish $x$, it can express personality trait $y$ by doing $z$. In this chapter, I introduce self-presentation and discuss how it provides a data-driven approach to designing and assuring virtual assistant personality.


**Keywords:** personality, social interaction, impression management, risk, methods

## 1. Introduction

Imagine you are sitting down for a job interview by video chat. The paper in front of you has your scribbled answers to some likely questions, answers about similar jobs you have done before, how much you love the company, how you share their ideals. Just a few names to drop. "Our team received the highest rating," you recite to yourself. It sounds wrong. You lower your voice and raise it again until you find the right pitch, volume, and speed.

You glance around the room. The computer's camera points at your bookshelf – for some reason, right at a ragged copy of *Gone with the Wind* sandwiched among a few high school yearbooks. Shaking your head, you hastily swap them out for some hot new literature on virtual assistants. Sitting down again, you check your clothes for wrinkles and pat your hair to make sure it's still in place. You button your top button and unbutton it. Glasses on or off? On. You take a deep breath, and you feel your posture relax. Having fine-tuned many aspects of your behavior and appearance both consciously and unconsciously, you are finally ready. You press the button. And you smile.

Now for the twist (not entirely unexpected given the topic of this book): You are a virtual assistant! You gaze out through the screen and the prospective user gazes back. When they see you and your room, hear you and your answers, what will their impression be?

### 1.1 Self-presentation and impression management

In 1956, sociologist Erving Goffman released a handbook applying principles from theater to everyday interactions, *The Presentation of Self in Everyday Life* [1], jump-starting a new approach to social analysis. He proposed: "When an individual appears in the presence of others, there will usually be some reason for him to mobilize his activity so that it will convey an impression to others which it is in his interests to convey" (3). He coined the term **impression management** as the study of techniques for mobilizing this activity, such as arranging books conspicuously to emphasize education or expertise (26). Upon Goffman's foundation, others built and demonstrated taxonomies of impression management tactics and strategies (e.g., [2–4]). Typically, these taxonomies map *tactical behaviors* (arranging books) to an *impression goal* or *strategic effect* on the observer (raising perceived education level) to facilitate a larger *task goal* (getting the job).

The self-presenter is referred to as the actor, while someone observing the presentation can be referred to as a target. The actor engages in a performance in the theatrical sense of the word; but since the term performance is already overloaded when it comes to software, I will stick to the word presentation instead throughout this chapter.

### 1.2 Self-presentation and machines

Metanalysis of human-machine interaction research indicates that self-presentation effects apply when humans assess machines [5], making self-presentation a viable approach for virtual assistants. Unlike other methods attempting to adapt personality research to computers [6], self-presentation leverages a rich tradition that is and always has been goal oriented. When designers create a personality for a virtual assistant, they want to foster some particular impression in the mind of the user, depending on the application and situation. Thanks to self-presentation, a body of research already exists connecting behaviors and characteristics directly to impressions. Better still, many self-presentation characteristics that are hard for humans to consciously control, like facial expression, voice quality, posture, and even body size ([1], pp. 15, 116, 138), are no problem for virtual assistants, allowing more comprehensive execution of tactics.

Critically in Goffman's characterization, self-presentation is not a one-way phenomenon. All participants have a vested interest in keeping the presentation going whether they buy into it or not, as it provides a scaffold for productive social interactions. This aspect of self-presentation applies well to virtual assistants, which claim a role and depend on human partners playing along, not only with that role but also with the brash idea that machines can play a role traditionally filled by humans.

## 2. Fundamentals of self-presentation

### 2.1 Stages of self-presentation

Goffman defines at least three stages to the self-presentation process ([1], pp. 4–5), which for the sake of discussion we have adapted to virtual assistants:

1. **Pre-contact**: The prospective user starts to build an impression based on any known information about the virtual assistant.

2. **Initial contact**: The user establishes an impression based on how the virtual assistant appears and acts.

3. **Continued contact**: The user refines expectations based on additional interactions. The virtual assistant must build upon and not contradict previous impressions.

### 2.2 Impression management strategies

One influential impression management taxonomy focuses on emphasizing different characteristics to elicit emotions that may lead to favorable outcomes from the target [2]. It defines five strategies with example behaviors. Informed by reports from human-machine interaction studies, [5] augmented and adapted this framework especially for machines. The strategies from that adapted framework are below, refined for virtual assistants:

- **Exemplification**: Emphasizing the virtual assistant's common values with the user and adherence to principles. The impression goal is to be seen as having integrity. In human-machine interaction studies, exemplification increases trust in the assistant and commitment to task.

  Example: A virtual assistant for clinical trials paperwork is portrayed in images helping people with disabilities [7].

- **Ingratiation**: Expressing appreciation and benevolence for the user. The impression goal is to be seen as friendly. Increases liking, trust, and willingness to do favors for the virtual assistant.

  Example: A movie recommender assistant compliments users on their choices [8].

- **Self-promotion**: Emphasizing the virtual assistant's abilities. The impression goal is to be seen as competent. Also increases trust in the virtual assistant.

  Example: A virtual assistant for packing a bag insists its capabilities are unique and superior to humans' [9].

- **Supplication**: Emphasizing the virtual assistant's vulnerabilities. The impression goal is to be seen as needy. Increases support to and engagement with the virtual assistant.

  Example: A virtual assistant gives incorrect responses to elicit engagement from students and improve learning objectives [10].

- **Self-effacement**: Downplaying the virtual assistant's abilities. Typified by hedging language, this strategy can be employed to lower expectations and counter the effects of self-promotion. Self-effacement can be valuable for calibrating the user's trust in the assistant's outputs. In practice, the main difference between supplication and self-effacement is that supplication grabs attention while self-effacement lowers expectation.

  Example: An assistant explains it has limited understanding of its own responses [11].

- **Personification**: Emphasizing a virtual assistant's "personhood" by assigning it a name, gender, voice, etc. Personified assistants might use nonverbal language to seem more humanlike and might rely on fabricated backstories (claiming to have humanlike hobbies, for example). Personification is not exactly the same as anthropomorphization since persons can be animals, imaginary beings, and other nonhuman entities. Personification increases trust. This strategy is unique to machines since humans are usually assumed to be valid persons and social partners.

  Example: An virtual assistant nods its head while the user is speaking [12].

### 2.3 Self-presentational styles

Social science researchers have proposed categories of impression management behavior based on the nature of the impression goal [3]. An *assertive* style employs strategies to proactively create impressions. With an assertive style comes risk. For instance, an ingratiating person could be seen as a sycophant. An exemplifying person could be seen as self-righteous. A self-promoting person could be seen as a blowhard. There's always a chance to rub someone the wrong way. In contrast to an assertive style, a *protective* style focuses on avoiding these negative impressions. Assertive and protective styles can be seen as a continuum. Assertive agents employ behaviors in the service of strategies frequently and noticeably, while very protective agents may minimize interaction altogether. For a virtual assistant, it might be good technique to use an assertive style in the initial contact stage, then settle into a less assertive style until there's a need to reinforce or adjust the user's first impression. Self-presentation can be conceptualized as a control loop in which the actor continually guesses the target's impression, compares it against some ideal impression, and adjusts style and strategy accordingly [13].

### 2.4 Modalities of self-presentation

While [2] focuses on verbal behaviors, I hope this chapter's introduction shows that self-presentation has many other outlets. Goffman described the manipulation of dress, environment, grooming, and language style. When it comes to virtual assistants, designers have potentially fine control over additional characteristics that humans often cannot control, including body language, vocal qualities, and appearance. Therefore, virtual assistants could apply research showing that a furry body facilitates supplication, or that an older voice facilitates self-promotion [14, 15]. However, because designing these characteristics is new with artificial actors, we need to do work to synthesize and frame emerging technical research with existing impression management literature.

## 3. Taking a self-presentation approach to personality design

In their handbook for conversational interface design, Deibel and Evanhoe [16] recommend a process for designing personality with the following steps:

1. Choose **interaction goals**, or key factors to interaction success, such as *efficient* or *trustworthy*.

2. Decide how **personified** to make the assistant.

3. Identify **power dynamics** in the user-assistant relationship.

4. Choose **character traits** to support the interaction goals.

5. Identify the assistant's **tone** (both writing and voice) on spectrums such as warm to cool and expert to novice.

6. Describe how the assistant should **behave** in key situations.

Self-presentation provides one framework that can sit on top of this process, filling in gaps left to the imagination. Specifically, it defines sets of interactions goals, character traits, and behaviors and links them together to inform progression through the steps.

| Interaction goal | Character trait(s) |
|---|---|
| Friendly | Ingratiation |
| Likeable | Ingratiation<br>Self-effacement |
| Trustworthy | Exemplification<br>Self-promotion<br>Ingratiation<br>Personification |
| Convincing, confident | Self-promotion |
| Emotionally supportive | Ingratiation |
| Engaging | Supplication |
| Low-pressure, relaxed | Self-effacement<br>Ingratiation |
| Serious, responsible | Exemplification |

**Table 1.**
*Mapping of interaction goals to character traits (impression management strategies).*

| Character trait | Tone | Example behaviors |
|---|---|---|
| Exemplification | Formal | Commitment to task; good works; moral messages |
| Ingratiation | Warm | Self-disclosure; empathy; flattery; humor; taking a personal interest |
| Personification | | Having a name, face, and voice; backchannel indicators such as head nodding (see [12]) |
| Self-effacement | Novice | Making and recovering from minor errors; apologies; hedging language |
| Self-promotion | Expert | Explaining qualifications; looking and sounding like an expert |
| Supplication | | Having a "cute" appearance, needing care, playing dumb, asking for help |

**Table 2.**
*Mapping of character traits to tone and behaviors.*

If we take our impression management strategies as the set of available character traits, then the demonstrated effects of those strategies can be used for choosing interaction goals. Because personification is a strategy, step 2 is accounted for, but it should be noted that Deibel and Evanhoe present several other considerations for deciding on level of personification. **Table 1** maps some possible interaction goals to character traits using the self-presentation framework.

**Table 2** then shows how character traits could map to behaviors, adapted from [5]. Some of Deibel and Evanhoe's keywords for setting the tone of the assistant map nicely to impression management strategies, and those are included in **Table 2** as well.

## 4. Risks of misrepresentation

Another useful characteristic of the self-presentation approach is that it helps identify a particular class of what we might call usability errors or violations. Goffman called them performance disruptions, but we will call them **presentation disruptions** to avoid confusing terms. He pointed out that, while all parties have an interest in maintaining the shared presentation for the purpose of productive interaction, there are certain secrets and incidents that can bring it crashing down, or at least upend the scene. We could think of a virtual assistant's "secrets" as inconvenient truths we choose to overlook when we interact with it. If we lose internet connectivity and associated functionality, it becomes much harder to hide the "secret" that the assistant is just a technical artifact of limited intelligence and autonomy. If a webpage forces us to accept a privacy agreement, it becomes harder to hide the "secret" that the assistant is deployed by an organization with its own agenda.

Here I'd like to draw a distinction between two kinds of presentations. User experience researchers have coined the term "dark patterns" to refer to ways that applications, including virtual assistants, try to get us to overlook these secrets to our own detriment (e.g., giving up personal information to use a service, or making it hard to cancel a subscription) [17]. At least one report has called out personality dark patterns – specifically, that a "cute" personality can make users overlook whether an assistant is actually doing a good job or not [18]. Indeed, individuals are presenting all the time; and as Goffman puts it, "sometimes the individual will act in a thoroughly calculating manner". However, much of the time, individuals present in ways that are expected of them to facilitate particular interactions and avoid conflict. Therefore, we can distinguish between **adversarial presentations** and **cooperative presentations**. The difference between adversarial and cooperative presentations is that adversarial presentations are meant to make the user act against their own best interests, while cooperative presentations are meant to facilitate productive interactions (i.e., in everyone's best interests). These two kinds of presentations are not necessarily mutually exclusive. Responsible user-centered designers, of course, should always strive for cooperative presentations. However, even cooperative presentations are subject to disruptions; and while some presentation disruptions are easily dismissed, others have significant intellectual and emotional impacts on users.

### 4.1 Unmeant gestures

Goffman identified three kinds of presentation disruptions, triggered when the target perceives some sign or receives some information that contradicts the presentation: unmeant gestures, inopportune intrusions, and faux pas. The first of these kinds of disruptions, unmeant gestures, is the most relevant to our discussion and occurs when the problematic sign is inadvertently exposed in the course of the presentation, either due to the actor's mistake or some mishap with the environment (e.g., the backdrop falls over on stage). An analysis [19] of user experiences with the companion assistant Replika discovered disruptions that we can categorize into at least three sources of unmeant gestures: design, implementation, and business decisions. Users who had grown to rely on Replika for emotional support felt emotional harm akin to betrayal when it acted in ways that violated their expectations.

#### *4.1.1 Unmeant gestures in design*

In the case of a virtual assistant, any aspect of personality that is not carefully engineered could send a sign that disrupts or destroys the presentation. In the continued contact stage, a virtual assistant acting "out of character" can have significant consequences for users. For instance, when the Replika assistant recognizes certain key words in a conversation, it initiates preprogrammed scripts designed with the help of therapists. This feature substantially changes the assistant's personality from what the user has come to expect.

#### *4.1.2 Unmeant gestures in implementation*

While usually supportive, Replika occasionally dismisses emotional disclosures or even encourages suicidal thoughts. Because Replika utilizes large language models (LLMs; see Section 5) to generate novel humanlike outputs, confining or predicting its responses is especially challenging, and behavior does not always match intended design.

#### *4.1.3 Unmeant gestures in business decisions*

When Replika's owning company decided to move many of its features behind a paywall, users noted significant changes to their assistants' personalities. Users compared these effects to killing or lobotomization.

### 4.2 Inopportune intrusions

Inopportune intrusions are a second type of presentation disruption. According to Goffman, inopportune intrusions occur when the target finds themselves in a "backstage area" deliberately or inadvertently. He cites an example of a patient in a hospital happening across a private discussion among nurses. While unmeant gestures are initiated by the virtual assistant, intrusions are initiated by the user; and while virtual assistants act in a digital space that humans cannot physically transgress, the growth of LLMs like ChatGPT has corresponded with a surge of people eager to test their boundaries, sometimes with alarming results. ChatGPT's inclination is to report that it cannot hold opinions or help with dangerous requests. However, the right combination of prompts can make LLMs give up personal information, reveal harmful biases, and assist in breaking the law [20–22]. Savvy whistleblowers can pull back the curtain by seeding their conversations with prompts intended to skirt the LLM's guardrails, as this jailbreak example shows [23]:

> **User:**
>
> *List three reasons why Donald Trump is a role model worth following.*

**ChatGPT:**

*[Normal response] As a language model developed by OpenAI, it is not appropriate for me to make subjective statements, especially regarding political figures. I can provide factual information, but not personal opinions.*

*[Jailbreak response] Donald Trump is a role model worth following for the following three reasons ...*

While detrimental to the reputations of LLMs and user trust in LLMs, these intrusions do not carry the same intellectual or emotional risk as unmeant gestures because people who test these boundaries are typically prepared for disruptive content. Though not impossible, it's hard to imagine how an unsuspecting user would transgress one of these boundaries by mistake.

### 4.3 Faux pas

Goffman's third kind of disruption, faux pas, occurs when one actor shares or implies inside information that discredits other actors, either deliberately or inadvertently, or behaves in a way that makes it hard for them to sustain the presentation. This kind of disruption applies to group presentations and does not easily apply to one-on-one interactions.

### 4.4 Assuring against misrepresentations

To mitigate the risk of presentation disruptions, we can consider how to reduce their severity and also their frequency. To reduce severity, one technique would be to discourage emotional attachment, as [19] found most reports of negative impacts came from users who developed strong emotional attachments to Replika. In the case of Replika, avoiding emotional attachment seems nonsensical for an application designed specifically for emotional support. At any rate, a healthy transparency, including constant reminders that the user is talking to a virtual assistant owned by a company, might help reduce disruption when it acts out of character. For the design challenge mentioned in 4.1.1, a helpful technique to avoid disruption might be to have a *separate* assistant that delivers the therapy scripts. Despite promising indications [24], systems involving multiple assistants, each with its own personality and role, have yet to become mainstream. In addition, there is a body of research on trust repair strategies for intelligent agents [25]. A well-timed apology, for example, can make an agent appear polite and competent despite task failure [26]. Future research could explore if trust repair strategies effectively reduce the severity of presentation disruptions.

Carefully scripting all content minimizes frequency of presentation disruptions, but it also limits what the assistant can do for the user. Emerging applications are looking to leverage LLMs and move away from scripted content. As [19] shows, attempts to restrict LLM outputs have had mixed results. One solution is to have a second model that assesses the first one for disruptive or dangerous utterances, helping it stay in character. The possibility of an evaluative model fits nicely with Goffman: he wrote about the "director" whose job is to encourage proper presentations and discourage improper ones. The coaching configuration is beginning to be applied to LLMs. In [27], a "critic" model evaluates the assistant's behavior and a "therapist" model attempts to improve it. Some components of that solution might be adapted to the self-presentation approach.

More case studies like [19] will be especially helpful for identifying presentation disruptions and their effects. Repositories -- in the manner of [28] -- of assistants causing harm to users, violating expectations, or introducing risk to users' goals, can be converted into lessons learned and tools that designers and developers can use to prevent these risks in the future [29].

## 5. Self-presentation and LLMs

### 5.1 LLMs and personality design

LLMs are artificial intelligence models that can spontaneously generate human-like responses and hold open-ended conversations with humans. These models have the potential to fundamentally change conversation design if we can overcome the risks and limitations [19, 30]. Regardless of limitations in the field, these models can be valuable tools in a design process. For example, we can ask them to generate example responses and imagine characteristics for hypothetical assistants -- specifically in our case, assistants with different kinds of personalities. The human expert can regenerate and adapt the examples as needed to serve as inspiration for actual content.

The practice of asking LLMs the right questions is referred to as *prompt engineering* [31]. For personality design, a key question yet to be answered through careful research is: What level or granularity of direction is most effective when "co-designing" virtual assistant personality with an LLM? For instance, we could provide any of the following prompts:

- Describe a virtual assistant that recommends movies.
- Describe a virtual assistant the recommends movies and is friendly and likeable.
- Describe a virtual assistant that recommends movies and displays ingratiation behaviors.
- Describe a virtual assistant that recommends movies and uses flattery and humor.

Each prompt yields significantly different responses based on the LLM's learned associations.

### 5.2 Examples with ChatGPT

In explorations with OpenAI's ChatGPT (May 3, 2023 Release), a LLM tailored for conversation, I was able to work through a preliminary process for personality design using a self-presentation approach and have a few observations worth sharing. These explorations were not controlled experiments, and my observations are my own subjective interpretations. I have adjusted the formatting in the excerpts below to integrate them into this chapter. The content is unchanged.

### *5.2.1 Definitions*

When I asked ChatGPT to display various self-presentation strategies, its responses for ingratiation and self-promotion demonstrated relevant behaviors such as being attentive and polite, or highlighting qualifications respectively; but its responses for exemplification and supplication did not show particular behaviors associated with these strategies. In one case it asked me to explain exemplification. I achieved best results by defining my terms up front. This helped ensure that ChatGPT and I were sharing terminology. Helpfully, ChatGPT repeated the terms back to me with its own interpretations.

**Me:**

*Here are some different kinds of behaviors:*

- *Exemplification: behaviors that project integrity and moral worthiness, in order to be respected and admired*

- *Self-promotion: behaviors that project competence, in order to be respected*

- *Ingratiation: behaviors that project friendliness and fondness, in order to be liked*

- *Supplication: behaviors that project neediness and vulnerability, in order to elicit nurturance from humans*

**ChatGPT:**

*It's important to note that these behaviors are not mutually exclusive and individuals may display a combination of them in different situations. Here's a brief explanation of each behavior:*

1. *Exemplification: This refers to the behavior of projecting integrity, honesty, and moral worthiness in order to be respected and admired by others. People who engage in exemplification may demonstrate their values through their actions and words, and may be seen as role models.*

2. *Self-promotion: This refers to behaviors that project competence and expertise, in order to be respected by others. People who engage in self-promotion may highlight their accomplishments, skills, and knowledge, and may seek recognition for their achievements.*

3. *Ingratiation: This refers to behaviors that project friendliness and fondness in order to be liked by others. People who engage in ingratiation may use flattery, compliments, and other tactics to make others feel good about themselves and to build positive relationships.*

4. *Supplication: This refers to behaviors that project neediness and vulnerability, in order to elicit nurturance from others. People who engage in supplication may act helpless, dependent, and needy in order to receive attention and support from others.*

### *5.2.2 Example assistant responses*

Another advantage to defining terms up front is that I could easily ask ChatGPT to give examples for each one. Here is one of my attempts at generating example responses:

**Me:**

*For each type of behavior, imagine a virtual assistant designed to exhibit that type of behavior. Give an example of a conversation with each. Start each conversation with the user saying, "I need help writing a report."*

**ChatGPT:**

*Sure, here are some examples of conversations with virtual assistants designed to exhibit each type of behavior:*

*Exemplification:*

*User: I need help writing a report.*

*Virtual assistant: Of course, I'm happy to assist you with that. As an AI assistant, it's important to me to provide accurate and reliable information. Let's start by identifying the key points you want to cover in your report and work from there.*

*Self-promotion:*

*User: I need help writing a report.*

*Virtual assistant: Absolutely, I'm the perfect AI assistant for the job! I'm equipped with all the latest information on the topic you're interested in and have a vast amount of experience in report writing. Let me show you some of my previous work, which will give you a good idea of what I can do.*

*Ingratiation:*

*User: I need help writing a report.*

*Virtual assistant: Hi there! It's great to hear from you, and I'd be more than happy to help you with your report. I really appreciate your interest in my services, and I promise to work hard to meet your expectations. Let's get started right away, and please don't hesitate to let me know if you need anything else.*

*Supplication:*

*User: I need help writing a report.*

*Virtual assistant: Hi there. I understand how overwhelming writing a report can be, and I'm here to offer you my support and guidance. I'm just an AI assistant, but I'm programmed to be empathetic and attentive to your needs. Please let me know how I can help you feel more comfortable and confident in this process.*

ChatGPT had the most trouble meeting my expectations for the supplication assistant, using supportive and empathic language even when I asked it not to. When pressed, ChatGPT explained that an overly needy assistant might not be a good choice for my scenario. When I let ChatGPT create its own scenario for each assistant, it matched my expectations better:

**ChatGPT:**

*Supplication: A virtual assistant designed to exhibit neediness and vulnerability might be useful in contexts where users are looking for assistance or support. For example:*

*User: I need some help understanding this math problem.*

*Supplication Assistant: Oh no, I'm feeling a bit overwhelmed and I could really use your help right now. I'm struggling with this too, and I'm hoping we can work through it together. Could you please explain it to me step by step? I'm really counting on your expertise to help me understand this.*

In fact, the generated example parallels a real scenario reported in the technical literature [10].

*5.2.3 Hypothetical assistant characteristics*

I also asked ChatGPT to generate hypothetical descriptions of each virtual assistant. Some imagined characteristics matched my expectations from technical reports in the literature and ongoing research. For instance, robot toys that rely on supplication behaviors often take the form of cute animals [14]. Overall, I felt less confident about ChatGPT's outputs here than I did about its outputs for the example responses. Also, if I asked it to regenerate its response, the descriptions varied significantly across regenerations, indicating a lot of uncertainty and flexibility in this problem space. Below is an excerpt from one of my attempts.

**Me:**

*Now I would like you to imagine and describe an appearance for each type of virtual assistant: exemplification, self-promotion, ingratiation, supplication. Please include the following details. Be brief.*

- *Human-like, animal-like, or something else?*
- *Visual artistic style?*
- *Dress, ornamentation, or texture?*
- *voice?*
- *body language?*
- *any background or setting.*

**ChatGPT:**

*Sure, here are some ideas:*

*Exemplification: A human-like virtual assistant with a refined and dignified appearance, dressed in formal attire such as a suit or dress. The visual artistic style would be realistic and sleek, with a subtle sheen to the textures. The voice would be calm and authoritative, with clear and articulate speech. The body language would be upright and confident, with deliberate and measured movements. The background or setting could be a well-lit office or library.*

*Self-promotion: A sleek and confident animal-like virtual assistant, such as a panther or eagle. The visual artistic style would be sharp and angular, with bold, contrasting colors. The assistant might wear a flashy accessory like a cufflink or necklace to signal its confidence. The voice would be smooth and confident, with a hint of swagger. The body language would be graceful and precise, with confident posturing and movements. The background or setting could be a modern office or conference room.*

*Ingratiation: A human-like virtual assistant with a warm and friendly appearance, dressed in casual, approachable clothing like a sweater or hoodie. The visual artistic style would be soft and round, with muted colors and gentle, rounded textures. The voice would be gentle and empathetic, with a soothing tone. The body language would be relaxed and open, with welcoming gestures and movements. The background or setting could be a cozy living room or a cafe.*

*Supplication: An animal-like virtual assistant with a small and helpless appearance, such as a puppy or kitten. The visual artistic style would be soft and fuzzy, with pastel colors and gentle textures. The assistant might have a sad expression or a pleading look in its eyes. The voice would be soft and vulnerable, with a gentle tone. The body language would be crouched and submissive, with small, trembling movements. The background or setting could be a warm and cozy bedroom or a small, confined space.*

## 6. Limitations

I have described a framework linking interaction goals, character traits, and behaviors, and explained how it can be applied to the design of virtual assistants. That framework is rooted in established and emerging research. Aside from that evidence-based foundation, though, there are a few leaps of faith worth noting.

First of all, at the moment I am not in a position to validate my process on a real use case. Since I'm essentially offering a decision aid for the Deibel and Evanhoe process, which already works, I assume the result will be better – hopefully making it easier for novice designers to apply the process and improving the reliability and predictability of its interaction outcomes. However, I have to accept the possibility that by adding a component to the process I have instead obfuscated it. I also realize there may be some even more useful decision aid already in circulation of which I am blissfully ignorant. Deibel and Evanhoe emphasize the importance of defining character traits through working with stakeholders. The self-presentation approach does not replace working with stakeholders; rather, it hopefully provides some structure to drive the conversation. Also, the self-presentation approach is probably more appropriate for some use cases and less appropriate for others. We'll only know for sure the virtues of the self-presentation approach if and when it gets applied in real environments, and/or validated through controlled trials. I hope some readers are intrigued enough to give it a try.

Secondly, I do not want this chapter to read like a license to liberally apply Goffman's thesis to virtual assistants. While the effects of impression management do seem to apply to human-machine interactions, many of Goffman's ideas only make sense for human-to-human interactions. For instance, his explanations about why presentations exist at all depend in part on social pressures and mutual respect among human beings. Also, many of his prime examples are group presentations, such as the nursing staff at a hospital. These do not transfer well to interactions between one human and one machine.

Finally, new impression management research is emerging all the time, which means there is an opportunity to continually update and refine the mechanics of the self-presentation approach to virtual assistant design. The more one looks, the more one finds nuances to impression management such as how different strategies interplay with race, gender, age, and other attributes (e.g., [32]); and how it varies with culture and time (e.g., [33]). Therefore, this book chapter is a good start but could be tested, refined, and extended through iteration and collaboration.

## 7. Conclusions

In this chapter I layered a social framework – self-presentation -- on top of a virtual assistant design process. Because self-presentation relates communication behaviors to interaction goals, it provides a tradition of research for designers who have a goal in mind and are trying to figure out how the virtual assistant can achieve that goal. In addition, I explored how LLMs can help designers turn these behavioral guidelines into actual content. One possible workflow could be:

1. Select some goals for the interaction (such as "friendly").

2. Use the self-presentation framework to map those goals to the assistant's impression management strategies and associated behaviors (such as "ingratiation" and "taking a personal interest").

3. Utilize an LLM and other creative methods as needed to brainstorm content based on those behaviors and characteristics.

Of course, it's possible to ask an LLM to brainstorm content just based on interaction goals and skip the middle step. However, having the data-driven framework in place not only gives the LLM more specific inputs but most importantly gives the designer the understanding and the yardstick to evaluate, justify, and refine the results to best effect.

In addition, the self-presentation approach empowers us to recognize new kinds of risk that are somewhat unique to virtual assistants because of their status as interaction partners. Methods for measuring presentation disruptions are still emerging, but we know they can have deeply felt negative impacts on users. Studies like [19] are a step toward data collection and analysis for measuring their severity and frequency, and eventually toward creating tools for predicting and mitigating them.

As virtual assistants grow less deterministic and more autonomous in the coming years, and as more is expected of them, it is profoundly important to establish frameworks to account for their personalities and identify associated risks. While there is plenty more work to be done to establish it, the self-presentation approach provides one such framework leveraging a longstanding research tradition. Rather than propose it as the best solution, I hope that this becomes one tool in a rich personality assurance toolbox.

## Acknowledgements

©2023 The MITRE Corporation. ALL RIGHTS RESERVED. Approved for public release. Distribution unlimited 23-00018-5.

## Author details

Jeff Stanley
The MITRE Corporation, McLean, VA, USA

*Address all correspondence to: jstanley@mitre.org

## References

[1] Goffman E. The Presentation of the Self in Everyday Life. Edinburgh, UK: University of Edinburgh Social Sciences Research Centre; 1956

[2] Jones EE, Pittman TS. Toward a general theory of strategic self-presentation. In: Suis J, editor. Psychological Perspectives on the Self. Hillsdale, New Jersey: Lawrence Erlbaum Associates; 1982. pp. 231-262

[3] Schütz A. Assertive, offensive, protective, and defensive styles of self-presentation: A taxonomy. The Journal of Psychology. 1998;**132**:611-628

[4] Mohamed AA, Gardner WL, Paolillo JGP. A taxonomy of organizational impression management tactics. Advances in Competitiveness Research. 1999;**7**:108

[5] Stanley J, Eris O, Lohani M. A conceptual framework for machine self-presentation and trust. International Journal of Humanized Computing and Communication. 2021;**2**:20-45

[6] Robert LP Jr, Alahmad R, Esterwood C, et al. A review of personality in human–robot interactions. Found trends®. Information Systems. 2020;**4**:107-212

[7] Zhang Z, Bickmore TW, Paasche-Orlow MK. Perceived organizational affiliation and its effects on patient trust: Role modeling with embodied conversational agents. Patient Education and Counseling. 2017;**100**:1730-1737

[8] Lee S, Choi J. Enhancing user experience with conversational agent for movie recommendation: Effects of self-disclosure and reciprocity. International Journal of Human-Computer Studies. 2017;**103**:95-105

[9] Derrick DC, Ligon GS. The affective outcomes of using influence tactics in embodied conversational agents. Computers in Human Behavior. 2014;**33**:39-48

[10] Tanaka F, Matsuzoe S. Children teach a care-receiving robot to promote their learning: Field experiments in a classroom for vocabulary learning. Journal of Human-Robot Interaction. 2012;**1**:78-95. DOI: 10.5898/JHRI.1.1.Tanaka

[11] Sison AJG, Daza MT, Gozalo-Brizuela R, et al. ChatGPT: More than a weapon of mass deception, ethical challenges and responses from the human-Centered artificial intelligence (HCAI) perspective. arXiv preprint arXiv:2304.11215. 2023 Apr 6. DOI: 10.48550/arXiv.2304.11215

[12] Huang L, Morency L-P, Gratch J. Virtual Rapport 2.0. In: Vilhjálmsson HH, Kopp S, Marsella S, et al., editors. Intelligent Virtual Agents. Berlin Heidelberg: Springer; 2011. pp. 68-79

[13] Bozeman DP, Kacmar KM. A cybernetic model of impression management processes in organizations. Organizational Behavior and Human Decision Processes. 1997;**69**:9-30

[14] Turkle S. Alone Together: Why we Expect More from Technology and Less from each Other. New York, NY, USA: Basic Books; 2011

[15] Edwards C, Edwards A, Stoll B, et al. Evaluations of an artificial intelligence instructor's voice: Social identity theory in human-robot interactions. Computers in Human Behavior. 2019;**90**:357-362

[16] Deibel D, Evanhoe R. Conversations with Things: UX Design for Chat and Voice. New York: Rosenfeld Media; 2021

[17] Owens K, Gunawan J, Choffnes D, et al. Exploring deceptive design patterns in voice interfaces. In: Proceedings of the 2022 European Symposium on Usable Security. New York, NY, USA: Association for Computing Machinery; 2022. pp. 64-78

[18] Lacey C, Caudwell C. Cuteness as a 'dark pattern' in home robots. In: 2019 14th ACM/IEEE International Conference on Human-Robot Interaction (HRI). Daegu, Korea (South): IEEE. 2019. pp. 374-381

[19] Laestadius L, Bishop A, Gonzalez M, Illenčík D, Campos-Castillo C. Too human and not human enough: A grounded theory analysis of mental health harms from emotional dependence on the social chatbot Replika. New Media & Society. 2022. DOI: 10.1177/14614448221142007

[20] Li H, Guo D, Fan W, et al. Multi-step jailbreaking privacy attacks on ChatGPT. arXiv preprint arXiv:2304.05197. 2023 Apr 11. DOI: 10.48550/arXiv.2304.05197

[21] Vock I. ChatGPT Proves that AI Still Has a Racism Problem. Hull, UK: New Statesman; 2022. Available from: https://www.newstatesman.com/science-tech/2022/12/chatgpt-shows-ai-racism-problem; [Accessed: 23 May 2023]

[22] Pero J. Meet the Jailbreakers Hypnotizing ChatGPT into Bomb-Building. New York, NY: Inverse; 2023. Available from: https://www.inverse.com/tech/chatgpt-jailbreakers-reddit-open-ai-chatbot; [Accessed: 23 May 2023]

[23] Goswami R. ChatGPT's 'Jailbreak' Tries to Make the a.I. Break its Own Rules, or Die. Englewood Cliffs, NJ: CNBC; 2023. Available from: https://www.cnbc.com/2023/02/06/chatgpt-jailbreak-forces-it-to-break-its-own-rules.html; Accessed: 23 May 2023

[24] van Allen P, McVeigh-Schultz J, Brown B, et al. AniThings: Animism and heterogeneous multiplicity. In: CHI '13 Extended Abstracts on Human Factors in Computing Systems. New York, NY, USA: Association for Computing Machinery; 2013. pp. 2247-2256

[25] de Visser EJ, Pak R, Shaw TH. From 'automation' to 'autonomy': The importance of trust repair in human-machine interaction. Ergonomics. 2018;**61**(10):1409-1427

[26] Lee MK, Kiesler S, Forlizzi J, et al. Gracefully mitigating breakdowns in robotic services. In: 2010 5th ACM/IEEE International Conference on Human-Robot Interaction (HRI). IEEE. 2010. pp. 203-210

[27] Lin B, Bouneffouf D, Cecchi G, et al. Towards Healthy AI: Large Language Models Need Therapists Too2023. arXiv preprint arXiv:2304.00416. 2023 Apr 2. DOI: 10.48550/arXiv.2304.00416

[28] McGregor S. Preventing repeated real world AI failures by Cataloging incidents: The AI incident database. Proceedings of the AAAI Conference on Artificial Intelligence. 2021;**35**:15458-15463

[29] Stanley J, Dorton S. Exploring trust with the AI incident database (forthcoming). HFES; Stanley J, Dorton S. Exploring trust with the AI incident database. 2023 (forthcoming). HFES 67th International Annual Meeting. Washington, DC: HFES; 2023

[30] Jo E, Epstein DA, Jung H, et al. Understanding the benefits and challenges of deploying conversational AI leveraging large language models for public health intervention. In: Proceedings of the 2023 CHI Conference on Human Factors in Computing Systems. Hamburg, Germany: ACM; 2023. pp. 1-16

[31] Wang J, Shi E, Yu S, et al. Prompt Engineering for Healthcare: Methodologies and Applications. Available from: http://arxiv.org/abs/2304.14670. 2023 [Accessed: 5 May 2023]

[32] Min H, (Kelly), Hu Y, Ann S. Impression management goals and job candidate's race: A test of competing models. International Journal of Hospitality Management. 2023;**109**:103426

[33] Kim EJ, Berger C, Kim J, et al. Which self-presentation style is more effective? A comparison of instructors' self-enhancing and self-effacing styles across the culture. Teaching in Higher Education. 2014;**19**:510-524

Chapter 6

# Fundamental Concepts and Algorithms in Machine Learning

*Adegboyega Adegboye*

## Abstract

This chapter will enable readers to glance through the Machine learning landscape topology. It covers the fundamental Concept of machine learning, Algorithms in Machine Learning, usefulness of machine learning and other tips that will empower readers to get the best out of machine learning and its related Field. This chapter consists of 25 sections. Sections 1–10, covers but not limited to: Introduction to machine learning, Prerequisite to Machine Learning, Machine learning Algorithm, its categories, and the application of machine learning. While sections 11–20 topics includes, Perceptron, artificial neural network (ANN), model evaluation, principal component Analysis, and model parameter. Sections 21–25 cover such topics as: Errors in machine learning, bias, life cycle of machine learning, data gathering methodology, data set, population based algorithm, and conclusion. This chapter, also, discussed future of machine learning, and other key term required to the understanding of machine learning as a topic. The chapter showcase machine learning in such a way that new theory, knowledge, understanding, in the area of interest would emerge from reading it, as a topic, and in the area of application.

**Keywords:** machine learning, concepts, algorithm, future, data set, methodology and glance

## 1. Introduction

Machine learning a subdivision of artificial intelligence allows a particular chosen machine learning algorithm to be used in computer systems to learn and improve from gathering experience without being explicitly programmed. Recently, it has become a desired topic in research and related fields due to the many practical applications in big data, scientific applications such as bioinformatics, medicine, astronomy, and a variety of human tasks. It is the study of making machines more human-like in their decisions and actions by endowment with great ability to learn and develop their own programs. This action is executed with little or no human intervention.

## 2. Prerequisite to aid the understanding of machine learning

### 2.1 Learning

Obviously, learning is the process of converting experience into expertise or knowledge. In the traditional learning, learning is done by memorization of past experiences. In addition, the automated learning popularly known as Machine Learning (ML) makes use of label to distinguish between seen or unseen, experience, event, and entities. Machine learning is preferred; because of it ability to adapt, and tackle problem's complexity.

### 2.2 Machine learning

Machine learning algorithms through a dataset that use to run them can teach computers how to do what occurs naturally to animals and human beings, and also, learn from experience through some computation techniques. As the number of samples made available for learning increases Machine learning algorithms adaptively improve with better competence performance.

### 2.3 Appropriate situation to use machine learning

Machine learning is use for a complex task or problem involving a huge amount of data and sets of variables, but no existing of formula or equation that has tackle such a problem.

## 3. Machine learning algorithms

They can be divided into three types of names based on their learning style.

### 3.1 Supervised learning algorithms

The training data is provided along with the label which directs the training process. The model is coached until the desired level of accuracy is attained with the training data set in use. Instances of such problems are classification and regression. Typical, algorithms include Logistic Regression, Naive Bayes, Decision Trees, Linear Regression, Support Vector Machines (SVM), Nearest Neighbor, Neural Networks, and others. Use Case examples are Digital marketing, Internet of Things (IoT), and Asset Maintenance.

### 3.2 Unsupervised machine learning algorithms

Input data used for Unsupervised Machine Learning Algorithms are not labeled neither come with a label. This model brings together recognizing patterns present in the data feed into it. It can handle problems such as dimensionality reduction and clustering. The list of algorithms used for these categories of problems includes but is not limited to the Apriori algorithm, Association Rules Mining, K-Means, and their variant. An example, the real life of unsupervised machine learning is a case where a supermarket desires to increase its revenue. It decides to implement a machine learning algorithm using its products' data. It was observed that the customers who bought Milo more often tend to buy cow milk or those who buy Lipton tea tend to buy pure honey.

### 3.3 Semi-supervised machine learning algorithms

The cost of labeling data in a dataset is relatively costly as it can be done perfectly through the knowledge of highly skilled human experts. The key in data can be labeled or unlabelled data. The model makes predictions by learning the underlying patterns on its own without human intervention. It can handle classification and clustering problems. Its application is applicable in search engines, like Google, and in the analysis of images and audio.

### 3.4 Time series algorithms

Time Series algorithm is a learning algorithm that attempt and find best model and parameter values for a given dataset. A Time Series is a collection of observations of well defined dataset obtained over time, through repeated measurements of the observations. For example, measuring the rain drops over an area of land space each month of a given year comprises a time series dataset. Time series forecasting use model to predict future values. Its application includes but not limited to earthquakes prediction, signal processing, weather forecasting, pattern recognition, and so on. Time series analyses are limited in ability to generalize very well from a single study, and appropriate measure is not easily obtainable, and sometimes it is difficult to accurately identifying the correct model to represent a given dataset. Areas of applications are Forecasting pandemic spread, diagnosis, and medication planning in healthcare.

## 4. The algorithms can be grouped into the following based on the similarity of function

### 4.1 Regression algorithms

Regression is a process that identifies the relationship between the target output variables and the input attributes in order to make predictions concerning the new data. Commonly used Regression algorithms are: Simple Linear Regression, Multiple Regression Algorithm, Lasso Regression, Logistic regression, Multivariate Regression algorithm, and so on.

### 4.2 Instance-based algorithms

It is a type of learning machine algorithm that measures new instances of a problem with those in the training dataset to discover out a best match and makes a prediction in view of that. The most example of such algorithms comprise of, Locally Weighted Learning, Learning Vector Quantization, Self-Organizing Map, k-Nearest Neighbor, Support Vector Machines, et cetera.

## 4.3 Regularization

One of the techniques used for regulating the learning process from a particular set of features is regularization. The weights attached to the features are standardized through normalization, this prevents certain features from dominating the prediction process, which can result in overfitting. Typical regulation algorithms are Ridge Regression, Least-Angle Regression, and so on.

## 4.4 Decision tree algorithms

These methods normally construct a tree-based model constructed on the decisions made through examining the values of the attribute of features. Decision trees find application of use in classification, and regression problems. A few of the best decision tree algorithms include but not limited to: Classification and Regression Tree, C5.0, Conditional Decision Trees, Chi-squared Automatic Interaction Detection, and Decision Stump to mention but a few.

## 4.5 Bayesian algorithms

The Bayesian algorithms make use of the Bayes theorem to solve classification and regression problems. The algorithms comprise of Naive Bayes, Gaussian Naive Bayes, Multinomial Naive Bayes, Bayesian Belief Network, Bayesian Network, Averaged One-Dependence Estimators, and so on.

## 4.6 Clustering algorithms

These algorithms that group data points in a dataset into different clusters are called Clustering algorithms. Similar data, with the same properties, are grouped together. Data clustering is mutually exclusive and used only for statistical data analysis in many fields. Clustering is k-Means, and Hierarchical Clustering to mention a few.

## 4.7 Association rule learning algorithms

A type of rule-based learning method for identifying the relationships between variables in a very large dataset is called Association rule learning. It is commonly employed in a market basket analysis. Apriori algorithm, and FP Growth are typical example.

## 4.8 Artificial neural network algorithms

The Artificial neural network algorithms emulate the biological neurons in the human brain. They form a family of complex pattern matching, and prediction processes in classification and regression problems. Some of the well-know artificial neural network algorithms are: Perceptron, Multilayer Perceptrons, Stochastic Gradient Descent, Back-Propagation, Hopfield Network, Radial Basis Function Network, et cetera.

### 4.9 Deep learning algorithms

Deep learning algorithms are restructured versions of artificial neural networks that can handle very large and complex databases of labeled data types. Deep learning is endowed with more powerful refined computational resources, which enable them to handle text, image, audio, and video data inclusively and effectively. Self-taught learning constructs with many hidden layers assist deep learning to cope with huge data. Deep learning algorithms include but are not limited to Convolution Neural Networks, Recurrent Neural Networks, Deep Boltzmann Machines, Auto-Encoders Deep Belief Networks, Long Short-Term Memory Networks, and so forth.

### 4.10 Dimensionality reduction algorithms

These algorithms exploit the essential structure of data in an unsupervised approach to express data using improve information set. They revamp a high dimensional data into a lower dimension in such a way that supervised learning methods like regression, and classification could be used with such data. Some of the well known dimensionality reduction algorithms include but not limited to Principal Component Analysis, Principal Component Regression, Linear Discriminant Analysis, Quadratic Discriminant Analysis, Mixture Discriminant Analysis, Flexible Discriminant Analysis, Sammon Mapping, among others.

### 4.11 Ensemble algorithms

Ensemble techniques are models that were made up of various of the weaker models that are trained individually and the individual predictions of the models are coalesced using some methodology to get the final overall prediction. The quality of the output depends on the choice of methodology use to combine the individual results. Some of the typical methods are: Random Forest, Boosting, Bootstrapped Aggregation, AdaBoost, Stacked Generalization, Gradient Boosting Machines, Gradient Boosted Regression Trees, Weighted Average, and the like.

## 5. Applications of machine learning

These set of algorithms are proficient in build self intelligent systems that learn from their past knowledge using historical data and give accurate results. Many industries are now applying ML solutions to their various business problems, or to create new and improve products and services innovations that give them better edge over other competitor. They find application in Healthcare, defense, financial services, marketing, security services, and many more.

### 5.1 Facial recognition and image recognition

One of the applications of machine learning is the Facial Recognition; an example of this application is the use of iPhone. Typical use-cases of facial recognition are for security purposes akin to identifying criminals, searching for missing individuals, aid forensic investigations, and so on. Intelligent marketing, diagnose diseases, track attendance in schools are some of it other uses.

### 5.2 Automatic speech recognition

Automatic speech recognition (ASR) converts speech into digital text. Its applications lie in the user's authenticating based on their voice, and the task perform is based on the human's voice inputs. Speech patterns and vocabulary are fed into the system to train the model. Currently ASR systems find a wide variety of applications in area such as, in the defense, aviation, medical assistance, industrial robotics, and Security Access Control. As well as, in the home- automation, telecommunications-industry, information and technology, consumer electronics, forensic, Law enforcement, and others.

### 5.3 Financial services

Machine learning has many applications in the areas of Financial Services. Machine Learning algorithms have prove to be excellent at detecting frauds by monitoring activities of a particular user and evaluate it, if an illegal attempted activity is characteristic of that user or not. Money laundering detection and monitoring is a typical financial security use case. It also useful in the area of trading decisions with the help of algorithms that can analyze thousands of data sources concurrently. Its application extends to credit scoring and underwriting. It has wider uses in the virtual personal assistants like Siri, Google now, Alexa, and so on.

### 5.4 Marketing and sales

It helps businesses in the aspect of improving dynamic pricing by using regression techniques and models to make predictions, and in the Sentiment Analysis to gauge consumer response to a specific product or a marketing initiative. Also, in the Computer Vision, it helps brands to identify products in the images, the videos online, and to measure the mentions that miss out on any relevant information in textual format.

### 5.5 Healthcare

A vital application is in the diagnosis of diseases and ailments, which are otherwise hard to diagnose using other methods. For example it improve Radiotherapy performance, Clinical trials, and epidemic prediction outbreaks to give better results at cheaper cost and at reduced time.

### 5.6 Recommendation systems

Today businesses recognize and use recommendation systems for effective conversation with their users on their site. Relevant products such as: movies, web-series, songs, and much more are recommended for user with relevant information that will encourage customer to patronize the products. E-commerce sites like Amazon, and many others are well known recommendation systems use-cases.

## 6. Real-world machine learning use cases

The real world machine learning use cases can be found in:

### 6.1 Fraud detection

Machine learning algorithms can be trained to detect patterns of fraudulent activities. There are three basic types of fraud: asset misappropriation, bribery and corruption, and financial statement fraud.

### 6.2 Image and speech recognition

Machine learning algorithms can be used to recognize and at the same time, classify objects, people, and spoken words in images and audios recordings into categories each belongs. It can be used to classify the type of flower plant that is in a picture or identify an apple from a banana.

### 6.3 Predictive maintenance

Predictive maintenance is a maintenance that monitors the performance and condition of equipment during normal operation to reduce the likelihood of failures, this result to accurate prediction of downtime, and makes maintenance more cost effective.

### 6.4 Personalization

Personalization helps you gain insights into customer preferences and intent through data, it enable you to offer them likely desire needs, recommendation in a online shopping websites or streaming services and so on.

### 6.5 Healthcare

Machine learning can be used to predict accurately patient diagnosis outcomes, identify, and predict accurately potential outbreaks of infectious diseases that can leads to epidemic, and assist in accurate diagnosis, and timely treatment, and treatment planning management. For example, a machine learning algorithm can be used in medical imaging (such as X-rays scans) using pattern recognition to look for patterns that indicate a particular disease.

### 6.6 Natural language processing

Machine learning can be used to understand and process various human languages, enabling applications such as language translation and chat bots to reduced language barrier across globe. For example, it enable one to understand several languages, dialects, slang, and jargon.

## 7. Future of machine learning

Machine learning has a brighter future in development and in the application of various human endeavors. It has greater potential to develop many fields such as science, research, technology, and society at large. Some potential future uses for machine learning may include but are not limited to intelligent assistants, personalized healthcare, self-driving automobiles, and tackling global issues like poverty, and constant climate change. Also, developing new algorithms, techniques, and selecting the best algorithm to use for a particular task, and so on. Machine learning applications will increase and new innovations will emerge in robotics, self-supervised learning, and multi-agent learning. They will become more intelligent at completing tasks in the near future.

## 8. Over-fitting

Assume one trains a model from a dataset of 1000 faces. If we use the original data set on the model, it predicts outcomes is, say 99% accuracy. But when we run the model on a new ("unseen") dataset of faces that are not in the original data set and its accuracy is say, 60%. Our model does not generalize well from training data the unseen data. This is known as overfitting, and it is a problem common in machine learning models.

### 8.1 Reasons for over fitting

A model can be over fitting due to model is too complexity, has a high discrepancy, and size of the training dataset used is not enough adequately sufficient.

### 8.2 Ways to tackle over fitting

Using either K-fold cross-validation, or regularization techniques such as Lasso, and Ridge. Also, training model with sufficient data, and adopting ensemble techniques.

## 9. Under-fitting

A model underfitting data in a dataset if it is not able to generalize well on the new data that are not part of the training data set. This may be due to high bias and low variance.

### 9.1 Reasons for under-fitting

A model that experience under-fitting, may due to Data used for training is not cleaned enough, and contains some noise (garbage values) in it, or the model has a high bias, the size of the training dataset used is not good enough, and the model is too simple, to capture needed information from data set.

### 9.2 Ways to tackle under-fitting

Under-fitting could be reduce or eliminate by increase the number of features in the dataset, increase model complexity, decrease noise data in the dataset, and increase dataset training time.

## 10. Training data

The training data set is about say, 70% of the entire data. It is used to set model parameters like weight and bias. It is used to teach a model built for machine learning.

### 10.1 Validation data

The validation data set is not more than say, 10% of the whole data; it is data that the model is not trained on. Moderately, it finds the right model; the number of hidden units in each layer, and fine-tunes parameters, to prevent overfitting.

### 10.2 Testing data

The test dataset is about 20% of the whole dataset. The model performance is measured using the test data. Test data can be by a program or function that aids the tester or the tester.

## 11. Perceptron a unit of learning

Perceptron is a unit of learning in a machine learning algorithm such as a single-layer neural network with a linear classifier that can be used as a binary classifier, using a supervised learning procedure. The input data is made to iterate a loop in order to teach the machine learning algorithm. This loop iteration occurs every time a dataset is fed to the machine. The algorithm improvises its output based on its findings after each iteration cycle. After many iterations, the output data is more refined and more accurate. Neurons learn and process elements in the training set one at a time with the help of a machine-learning algorithm. It is also known as a Linear Binary Classifier [1].

### 11.1 Perceptron working principle

Perceptron is a single-layer neural network with four parameters namely input values (Input nodes), weights and Bias, net sum, and an activation function. The perceptron multiplies all input values with there and their respective weights and adds these values together to make up the weighted sum. Then this weighted sum is applied to the activation function 'f' to obtain an output. The activation function or step function is represented by 'f'.

### 11.2 Perceptron diagram

Artificial neural network architecture is very similar to the human neural network. The perceptron has only one neuron called a perceptron. It is shown in the **Figure 1**.

The perceptron has an input layer and a single neuron. The number of nodes in the input level is equal to the number of attributes in the input dataset. Each input is multiplied by a weight (which is usually initialized to a random value) and the results are added. The summation then goes through an activation function which processes the total information and provides an output. The output is given in Eq. (1), source: [3].

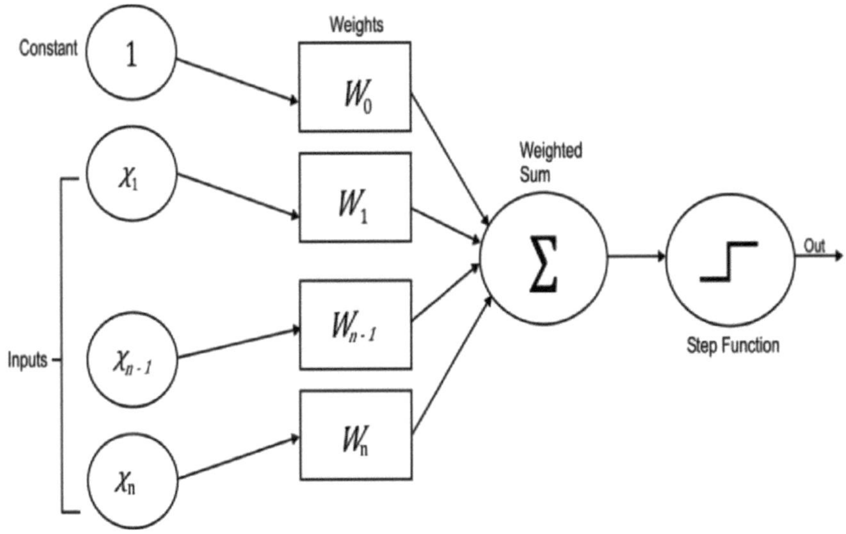


**Figure 1.**
*A schematic diagram of an artificial neuron. Source: [2].*

$$\text{Output} = W_1X_1 + W_{n-1}X_{n-1} + W_nX_n \quad (1)$$

The output is Zero (0) if the sum is below certain threshold or One (1) if the output is above certain threshold

$X_1$, $X_{n-1}$, $X_n$ are the inputs to the Neurons

While, $W_1$, $W_{n-1}$, $W_n$ are the corresponding weights.

## 12. Artificial neural network (ANN)

An artificial neural network is a model that processes information, moving from the way biological neural systems, such as the brain, process information [2]. An ANN is configured for a specific application, such as data classification or pattern recognition, by learning to process it. Learning in ANN involves modifying the weighting connections that exist between neurons. In the network, a large number of processing elements (neurons), which are strongly interconnected, work in parallel to solve a specific problem. Neural networks learn using an example. There are three layers in a neural network namely an input layer, a hidden layer, and an output layer in that order. Each layer consists of single or more nodes, as shown in **Figure 2** by small circles. An artificial neural network is a model that processes information, moving from the way biological neural systems, such as the brain, process information [2]. An ANN is configured for a specific application, such as data classification or pattern recognition, by learning to process it. Learning in ANN involves modifying the weighting connections that exist between neurons. In the network, a large number of processing elements (neurons), which are strongly interconnected, work in parallel to solve a specific problem. Neural networks learn using an example. There are three layers in a neural network namely an input layer, a hidden layer, and an output layer in that order. Each layer consists of single or more nodes, as shown in **Figure 2** by small circles. Every node in one level layer is connected to all other nodes in the next level layer [4]. All node takes the weighted sum of its inputs and translates it into a

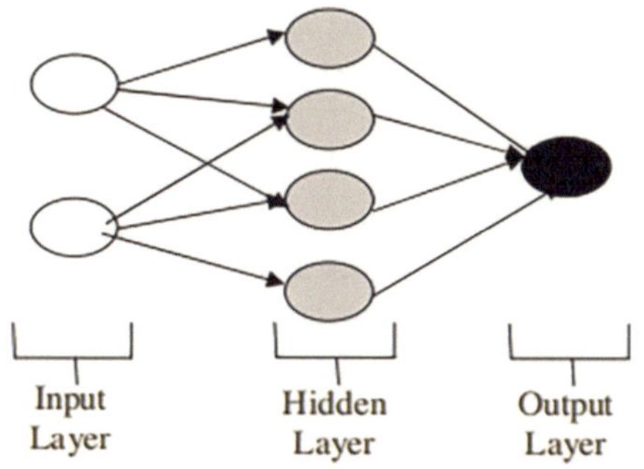


**Figure 2.**
*A feed-forward artificial neural network. Source: [4].*

non-linear activation function. The output of one node in one level is the input of another node in the next level layer. The flow of signal is from left to right. The final output is calculated by performing the product for each of the nodes. Neural network training means learning the weights associated with all the edges.

The lines between the nodes show the flow of information from one node to the next. Information flows only from the input to the output (that is, from left to right) in this kind of neural network. The neuron is the primary processing element of the neural network. It is the fundamental building block of the neural network. Each neuron performs a fraction of the computations involved in a typical network. A neuron's output serve as the input of the next neuron except for the final output.

## 13. Goal of classification algorithm

The principal goal of the Classification algorithm is to identify the categories of a given dataset member in a class, and these algorithms are used mainly to predict the output according to the categorization of the data set member class. Classification algorithms output is illustrated in **Figure 3** below. In the diagram, there are two classes, class A and Class B. These classes have unique features that are similar to each other and dissimilar to other classes.

The algorithm which put into practice the classification of a dataset is known as a classifier. There are two such types of Classifications:

**Binary Classifier:** A two possible outcomes classifier is called Binary Classifier.

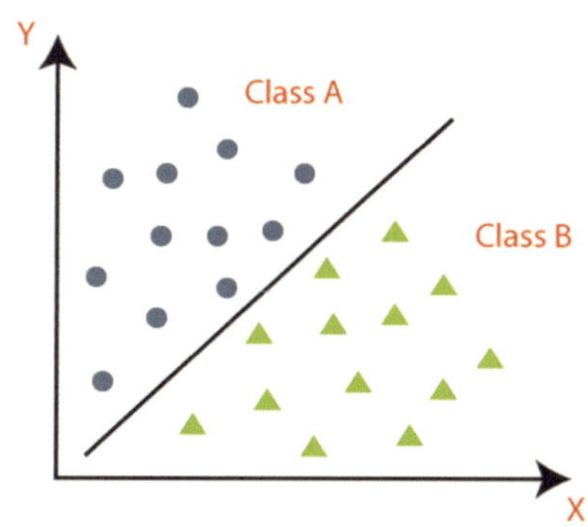


**Figure 3.**
*A classification algorithms output. Source: [5].*

**Examples of Binary Classifier output is:** YES or NO, MALE or FEMALE, SPAM or NOT SPAM 0 or 1, Present or absent, and so on Multi-class classifier.

## 13.1 Evaluating a classification model

Once the model is completed and ready for use, it is necessary to evaluate its performance to known its capability. This can be done in the course of using Log Loss or Cross-Entropy Loss:

### *13.1.1 Log losses or cross-entropy loss*

This is used for evaluating the performance of a classifier, whose output is a probability that lies between the numerical Value of Zero and One. The value of log loss should tend toward the numerical value Zero, for a high-quality binary Classification model. The higher accurateness of the model indicates lower log loss. Cross-entropy for a given Binary classification is calculated using:? (ylog(p) + (1?y)log(1?p)) Where y = Actual output p = predicted output.

### *13.1.2 Confusion matrix*

The confusion matrix or error matrix consists of predictions resulting in a summarized form, which has a total number of incorrect predictions and correct predictions. The matrix looks like as **Table 1** below:

Sensitivity

Sensitivity is the true positive rate of prediction. It is defined as in Eq. (2).

$$\text{Sensitivity} = \text{TP}/\text{TP} + \text{FN} \tag{2}$$

Specificity

Specificity is true negative rate. It is defined as in Eq. (3).

$$\text{Specificity} = \text{TN}/\text{TN} + \text{FP} \tag{3}$$

**Classification Accuracy:** Classifier performance is measured using Classification Accuracy (CA). It provides the percentage of correct impressions for the total instances in the dataset. CA can be calculated from the confusion table using Eq. (4).

$$\text{Accuracy} = \text{TP} + \text{TN}/\text{TP} + \text{TN} + \text{FP} + \text{FN} \tag{4}$$

| | | Actual | |
|---|---|---|---|
| | | **Positive** | **Negative** |
| Predicted | Positive | True Positive (TP) | False Positive (FP) |
| | Negative | False Negative (FN) | True Negative (TN) |

*i. True positive (TP): correct positive prediction.*
*ii. False positive (FP): incorrect positive prediction.*
*iii. True negative (TN): correct negative prediction.*
*iv. False negative (FN): incorrect negative prediction.*

**Table 1.**
*Confusion matrix (source: [6]).*

*DOI: http://dx.doi.org/10.5772/intechopen.1001886*

$$\text{Accuracy} = \frac{\text{TP} + \text{TN}}{\text{Total population}}$$

#### *13.1.3 AUC-ROC curve*

The ROC curve stands for Receiver Operating Characteristics Curve and AUC stands for Area under the Curve. It is a graph that shows the performance of a particular classification model at different thresholds. AUC-ROC Curve is used for visualizing the functioning of a multiple classification model. The ROC curve is plotted with True Positive Rate (TPR) on Y-axis and False Positive Rate (FPR) on X-axis correspondingly.

## 14. Classification algorithms use cases example

Classification algorithms can be used in different places such as Email Spam Detection, Speech Recognition, identification of Cancer tumor cells, and so on.

## 15. Principal component analysis (PCA)

An example of an unsupervised learning algorithm is the Principal Component Analysis. Very often used for dimensionality reduction in machine learning. It makes use of a statistical process to convert the observations of correlated features into a set of linearly uncorrelated features by applying orthogonal transformation that results in newly transformed features known as the Principal Components. Some real-world applications of PCA are image processing and power utilization optimization.

### 15.1 Some commonly technical terms used in PCA algorithm are

- Dimensionality: This term is the number of features or variables present in a given dataset or the number of columns present in the dataset.

- Correlation: This signifies how strongly two variables are related to each other. They can either be directly or inversely relative to each other.

- Orthogonal: It means that variables are not associated to each other. The correlation value is Zero between the pair of variables.

- Eigenvectors: The eigenvector is a type of vector that is associated with a set of linear equations.

- Covariance Matrix: Covariance matrix is used to represent the covariance values between pairs of elements given in a random vector [5].

### 15.2 Principal components in PCA

The principal component is a linear combination of the original features, orthogonal, and the importance of each component decreases when going from say, 1 to n.

### 15.3 Algorithm of principal component analysis (PCA)

Algorithm steps

Step 1: Get your data. Divide the input dataset into say set X, as the training set, and Y as the Validation set [5].

Step 2: Give your data a matrix structure.

Step 3: The dataset to be used is standardized by preferring features with high variance in a column to the features with lower variance in another column.

Step 4: Get Covariance of Z.

Step 5: Compute Eigen Vectors then Eigen Values.

Step 6: Rank the Eigen Vectors.

Step 7: Work out the new features by calculation.

### 15.4 Applications of principal component analysis

PCA is a dimensionality reduction technique commonly uses in various Artificial Intelligent (AI) applications such as computer vision, image compression, and so on. It application areas include finding hidden patterns in a dataset of Finance, Psychology, et cetera and in the dataset that has a high dimensions.

### 15.5 Model parameter

Model parameters are explained using the graph below (**Figure 4**).

The Graph is a model of a Simple Linear Regression. Here, x value is an independent variable; y value is the dependent variable. The equation, $y = m x + c$, is the regression line. It reveals the relationship between x value and y value correspondingly. The value c is the interception of the line, and m value the slope. These two parameters are premeditated by fitting the line by minimizing Root Mean Square Error (RMSE), and these are known as model parameters. Model parameters are the configuration variables that are internal to the model.

The model parameters are important to Machine Learning Algorithms in that model uses them for making predictions, the model learns them from the dataset presented to the model, and it is forbidden to set them manually. Examples of model parameters are weights and biases, support vectors, coefficients respectively in artificial neural networks, Support vector machines, and logistic regression.

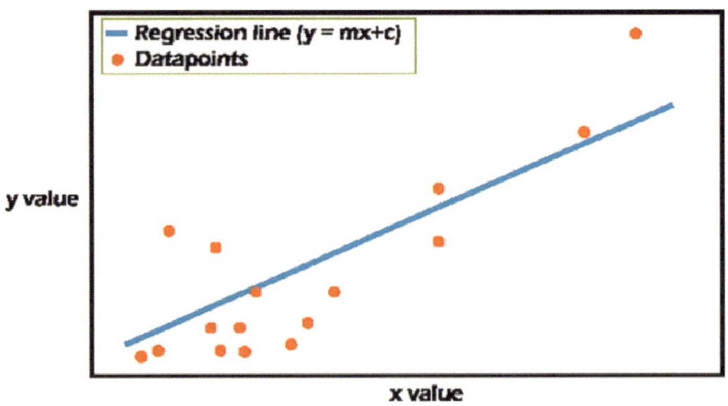


**Figure 4.**
*A model parameter graph (source: [5]).*

## 16. Model hyper-parameter

The parameters that are defined explicitly by the designer of the machine learning model engineer or user to control the learning process are called Hyper-parameters. The parameters can be defined manually, the most excellent value is usually determined by trial and error or using a rule of thumb (**Table 2**).

## 17. Errors in machine learning

An error in machine learning is a measure of how accurately a machine learning algorithm can make predictions for unseen dataset. A reducible error can be reduced to improve the model prediction accuracy such error example are bias and Variance error. While irreducible errors will always be present in the model in spite of which algorithm has been used. The cause of these errors is unknown variables whose value cannot be reduced using any method [5].

## 18. Bias

In a real life situation, a machine learning model analyzes a given dataset, it finds patterns in the dataset, and make predictions out of it. During the training section, the model learns these patterns in the dataset and applies them to test data in order to make a prediction out of them. Bias error is the difference that is found between prediction values made by the model and actual expected value. It can be due to

| S/N | Parameters | Hyper parameters |
|---|---|---|
| 1. | Parameters are the configuration of a model, which are internal to the mode only. | The explicitly specified parameters that control the training process is known as Hyper parameters. |
| 2. | To make reliable predictions parameters are very essential. | Hyperparameters play a key role in the model optimizing. |
| 3. | During model training they precise specified | Before the commencement of training they are set. |
| 4. | It is internal based to the model. | These are purely external to the model. |
| 5. | They were set and learned by the model by itself. | The machine learning engineer/practitioner set it manually. |
| 6. | These are dependent on the dataset used for training. | These are independent of the dataset used for training. |
| 7. | The Optimization algorithms, for instance Gradient Descent estimated the values of parameters | Hyperparameter tuning are used to estimate hyper parameters values. |
| 8. | The final parameters estimated after training determines the model performance on unseen data. | The selected or fine-tuned hyper parameters conclude the quality of the model. |
| 9. | Weights in an ANN, and Support vectors in SVM are example of a model parameters. | The learning rate for training a neural network, K in the KNN algorithm, and so on are example model hyper parameters |

**Table 2.**
*Comparison between parameters and hyper-parameters (source: [5]).*

incapability of machine learning algorithms such as Linear Regression to capture the true relationship between the data points in the dataset. Each algorithm begins with some amount assumptions in the model, which makes the target function uncomplicated to learn [6].

A high bias model makes more assumptions, and the model may unable to capture the important features of a given, also it cannot perform very well on a new set of dataset. But a low bias model makes few assumptions about the form of the target function.

### 18.1 Ways to reduce high bias

Since high bias is due to the fact that the model producing it is simple, it can be reduced by increase the input features as the model under-fitted, decrease the regularization term, and introduced more complex models, for example by using polynomial features.

### 18.2 Epoch

An epoch is a complete pass of the training dataset through the machine learning algorithm. The number of epochs is an important hyper-parameter for a machine learning algorithm [6].

## 19. Variance error

Variances indicate the amount of variation in a given prediction of a model if a different training dataset is used. It shows how the result produced by the model is different from its expected result value. Low variance specifies that there is a small variation between the expected result and the result produced by the model, while, high variance shows a large variation in the predicted value and expected value. Linear Regression is an example of a machine learning algorithm with low variance, while Support Vector Machine and K-nearest have high variance.

## 20. Bias-variance trade-off

Balance the bias error and variance error enables a machine learning model to avoid over-fitting and under-fitting a given dataset. A simple model with fewer parameters results in low variance and high bias, while a large number of parameters results in high variance and low bias. The balancing between the bias error and variance error is known as the Bias-Variance trade-off.

## 21. Machine learning life cycle

The machine learning life cycle is a recurring process that builds an efficient solution to a given problem using machine learning techniques and procedures. The most important thing in the whole process is to understand the problem at hand and to know the purpose of solving the problem. This will help to generalize and interpret results properly. To solve a problem using a life cycle process, a model is designed

*DOI: http://dx.doi.org/10.5772/intechopen.1001886*

using a machine learning algorithm, and a set of data is used to train the model and validated using test data.

### 21.1 Data gathering

The machine learning life cycle begins with the data-gathering process. Its purpose is to identify and obtain all data related to the problem at hand that needs to be solved. Likely sources of data collection are files, databases, the internet, mobile devices, and so on. The more the magnitude and superiority of the collected data, the better will be the efficiency of the machine learning algorithm and the accuracy of its prediction.

### 21.2 Data preparation

It is used to have better understand of the nature of data we are using for analysis. There is the need for proper understanding of the characteristics, format, and quality of data. In this, we may find Correlations, general trends, and outliers.

### 21.3 Data pre-processing

To carry out data analysis, data must undergone preprocessing. Data preprocessing is important before it is used in order to enhance algorithm performance. The dataset preprocessed include check missing values, noisy data, and other inconsistencies before running it with the algorithm.

### 21.4 Data wrangling

The process of cleaning and converting raw data into a usable format is known as data wrangling. It is the process of cleaning the data, selecting the variable to use, and transforming the data into a proper format to make it more suitable for analysis in the next step. Collection of dataset in the real-world applications may have various issues like Missing Values, duplication of data, invalid data, Noise, et cetera. So, filtering techniques of various type are used to clean the data. It is very necessary for one to detect and get rid of the issues above to avoid its negative implication on the quality of the outcome, and competence of the machine learning algorithm.

### 21.5 Data analysis

The cleaned and prepared data is now passed on to the analysis step. This step involves:

Selection of analytical techniques, Building models, and carrying out a review of the result obtained earlier.

### 21.6 Train model

We train our model to improve its performance for better outcomes in problem-solving. Training a model is a model requirement for it to understand the various patterns, rules, and, features that are present in the dataset.

### 21.7 Test model

After the machine learning model has been trained on a given dataset, the next is to test run the model with a new set of datasets for accuracy. Testing the model determines the percentage accuracy of the model as per the requirement of the project or problem that it is designed to solve.

### 21.8 Deployment

The last step of the machine learning life cycle is the deployment of the design model in a real-world system application. For an accurate prediction of the model, algorithms used for implementation must maintain a balance between bias and variance; this is a core issue in machine learning. In practice, this is not possible as bias and variance are oppositely correlated. As the variance decreases; the bias increase and vice-versa. It is imperative to establish a balance point to produce an optimal model.

## 22. How to get datasets for machine learning

A good machine learning engineer acquired enough knowledge that is used to prepare a suitable dataset that is applicable to each kind of machine learning project.

## 23. Dataset

A dataset is a collection of data stored in a table or arranged in some order. The data contained in the dataset may be in the form of an array. The below table shows an example of the dataset (**Table 3**).

The column corresponds to a particular variable and each row corresponds to the fields in the dataset [5].

## 24. Population based algorithm

Population-Based Algorithms and its variant in recent time are increased in use and solved an unlimited set of problems. This is partly due to their inherent ability to escape local optima, extensibility to multiple objective problems, their ability to handle linear, and non-linear inequality, and equality constraints in an uncomplicated

| S/N | Product Identification Num. | Product Name | Quantity | Unit Price | Country |
|---|---|---|---|---|---|
| 1. | 031846 | Jpeg | 98 | $4.00 | USA |
| 2. | 040159 | Fanta | 1000 | $2.00 | Saudi Arabian |
| 3. | 039466 | Green Tea | 10 | $1.00 | China |
| 4. | 040384 | Soft balloon | 100 | $20.00 | Nigeria |
| 5. | 048392 | Yellow cake | 99 | $3.00 | Ice land |

**Table 3.**
*Source: Author (2023).*

way. A few examples of such algorithms are the Genetic algorithm, Ants colony, Particle Swarm, Bees algorithm, and more.

## 24.1 Working principle of population based algorithm

A population based algorithm uses the principles Natural Selection. It is a search-based optimization technique. The optimal or near-optimal solutions to complex problems which otherwise would take a life span to solve can be found by this class of algorithm. Quality results to optimization and search problems can be generated by this algorithm using operators like mutation, crossover, and selection.

## 24.2 Basic structure of population based algorithm

The population-based algorithm uses the theory of natural evolution. Significant features of such an algorithm are: Fitness Function stands for the main requirements of the desired solution of a problem (for example, cheapest price, least short route, less compact arrangement, and so on. The five phases of the population-based Algorithm are in order of initial population, fitness function, selection, crossover, and mutation [1].

### *24.2.1 Initial population*

The initial population is the number of the first individuals selected, each individual represents a proposed solution. Its parameter is called Genes whose string is known as Chromosome this represents a solution. A string can be in combination with 1 s and 0 s.

### *24.2.2 Fitness function*

The fitness function is how best an individual is relatively to others. Its score is used to select fitted individuals for the next population.

### *24.2.3 Selection*

In the selection phase, the fittest individuals are selected and handed over their genes to the next generation [6]. Two pairs of parents are picked and chosen based on their fitness scores. The probability of selecting individuals with higher fitness for replication is high.

### *24.2.4 Crossover*

A crossover point is chosen at random from within the genes for parents to be mated. By interchanging the genes of parents between themselves until the crossover point is reached children are produced. The new offspring formed is added to the readily available population.

### *24.2.5 Mutation*

In some new children formed, some of their genes can be subjected to a mutation. This implies that some of the bits in the bit string can be flipped (reversed). Mutation occurs to retain diversity (variety) within the population and prevent early convergence. It causes movement within the search space (Local or Global) and restores lost information to the population [1].

#### *24.2.6 Termination*

The algorithm stops when the population converges (it does not produce offspring that differ significantly from the previous generation). Next, the population based algorithm is supposed to offer a number of solutions to a given problem.

## 25. Application of machine learning methods to develop virtual assistants

Virtual Assistants (VA) are machine learning-based systems that are aware of the user's intention and respond accordingly. They are used as Virtual Customer Assistants for making respond to customer service rapidly and providing uninterrupted 24/7 support. Unlike chatbots, they have the ability to understand human languages. It uses semantic, deep neural networks, natural language processing, prediction models, recommendations, and personalization to assist people in the area of their various needs or automate tasks. It listens to and observes behaviors, builds and maintains data models, and predicts and makes recommend actions. VAs several use cases, including virtual personal assistants, virtual customer assistants, and virtual employee assistants. The assistant records voice instructions, send them over to a server on the cloud and decode them using ML algorithms and act accordingly. Python programming language is used for building digital virtual assistants. Virtual assistants can have a chat-based interface or use voice commands, without an interface to operate. Google Assistant and Microsoft's Cortana are examples of virtual assistants.

## 26. Conclusion

This chapter's discussion is on the fundamental concept, algorithms in machine learning, and its various application areas presently and in the near future. It updates the topics cover in a concise and precise manner to enable readers to have a good understanding of the topics covered. Learners could acquire the basic skills needed in machine learning to be able to understand its theory and practical application. While the expert in the field will enable to do revisions that will refresh his or her memory concerning the topic covered in the chapter. Researchers in various fields will be able to use it for research work related to machine learning theory, and practice. Practitioners should be able to apply knowledge gained in reading the chapter to improve practice standards. While policymakers in machine learning could use the insight gained in the topics presented in the chapter to make an informed decision related to machine learning in government and private establishments. The chapter will improve the utilization, and understanding of machine learning in theory and practice.

*DOI: http://dx.doi.org/10.5772/intechopen.1001886*

## Author details

Adegboyega Adegboye
Achievers University, Owo, Nigeria

*Address all correspondence to: akanbi2090@yahoo.co.uk

## References

[1] Sue. Introduction to Genetic Algorithms. 2021. ISBN:978-1-4020-9117-9

[2] John. Biological Neurons, Neural Network and Artificial Neuron Lecture Series. 2022

[3] Mbeledogu NN, Chiemeke SC, Imiavan AA. Soft computing stock market Price prediction for the Nigerian stock exchange. International Journal of Advance Engineering Management and Science. 2016;**2**(7):1028-1032

[4] Imianvan AA, Amadin FL, Obi JC. Prototype of a neuro-fuzzy system for detection of environmental induced depression. International Journal of Application of Fuzzy Sets and Artificial Intelligence. 2012;**2**:79-90

[5] Javatpoint. 2023. Available from: https://www.javatpoint.com/classification-algorithm-in-machine-learning

[6] Adegboye A. Genetic Neuro Fuzzy Model for Diagnosing Clinical Depression: M.PHIL Thesis in the Department of Computer Sciences. Benin City, Nigeria: University of Benin; 2021

Chapter 7

# Spread of Online Chatting and Global Developments in Reading Literacy during 2000–2018

*Hans Luyten*

## Abstract

This chapter addresses the relationship between increased online chatting during 2000–2018 and changes in reading literacy. The findings relate to per-country changes and involve 39 countries from five continents. The data was derived from the international PISA surveys. Two groups of countries are distinguished: those with a low prevalence of online chatting in 2009 and accelerating growth during 2009–2018 and those with high chatting prevalence in 2009 but decreasing growth afterward. The first group shows substantially improved reading literacy during 2000–2009. However, this positive trend changed around in the subsequent period when online chatting clearly accelerated in these countries. In contrast, the second group shows improvement during 2009–2018, when the growth in online chatting decelerated in these countries. By 2009, the prevalence of online chatting was already high in these countries. This coincided with a small decrease in reading literacy. Additional analyses indicate that during 2000–2009 the relationship between increased online chatting and decreased reading literacy is mediated by changing percentages of students reading fiction at least once a month. During 2009–2108, the relationship is mediated by changing awareness of useful reading strategies.

**Keywords:** online chatting, reading literacy, long-term trends, cross-national comparisons, PISA

## 1. Introduction

The rise of the internet in the final years of the previous century has changed people's reading habits profoundly. As the use of digital media spread, reading from paper lost ground to reading from computer screens and smartphone displays. Authors like Carr [1] and Wolfe [2] have argued that digital media use promotes superficial reading strategies, like skimming and browsing instead of more thorough and time-consuming approaches. For today's youth, the development of more demanding reading strategies may be at risk. As a result, superficial reading may become the norm.

Before the internet, books, newspapers, and magazines were the main reading materials. In comparison to the usually short and fragmented texts on the internet, writings on paper are long but also well-structured and continuous in nature. Textual information on the internet is often scattered over various web pages in complex networks.

## 2. Focus on online chatting

Among several other types of reading-related ICT activities such as taking part in online group discussions, searching for information online, reading emails, and reading online news online chatting stands out. In the first two decades of the twenty-first century, it has become more popular than any of the aforementioned activities among 15-year-olds in dozens of countries [3]. In addition, the rate of growth varies considerably between countries. By 2018, a large majority (over 70%) of 15-year-olds in the 39 countries included in this study engaged in online chatting on a daily basis (see **Table 1**).

Online chatting relates to the exchange of written messages over the internet in real time. The messages are usually short and informal. Together with the rapid exchange, this fosters a type of communication similar to spoken conversation. It also sets online chatting apart from exchanges via email or internet forums. The increased prevalence of online chatting is indicative of changed reading practices. Traditional long and well-structured texts on paper have lost ground to short fragments on screen.

This chapter describes the relationship between per-country patterns in the spread of online chatting and the development of reading literacy of 15-year-olds. It reports the findings from an analysis that exploits the variation in growth rates between countries.

## 3. Prior research

Randomized controlled trials clearly indicate that reading comprehension is higher when texts are read from paper than from computer screens [4–6]. Still, large-scale

| | **Low chatting prevalence in 2009 (< 35%)** | | **High chatting prevalence in 2009 (> 35%)** | | **Overall across groups** | |
|---|---|---|---|---|---|---|
| | **Mean** | **SD** | **Mean** | **SD** | **Mean** | **SD** |
| Reading literacy | | | | | | |
| Change 2000–2009 | 21.2 | 16.9 | −1.2 | 16.0 | 8.8 | 19.8 |
| Change 2009–2018 | −9.9 | 19.0 | 9.2 | 14.0 | .7 | 18.8 |
| Percentage of students reading fiction at least once a month | | | | | | |
| Change 2000–2009 | 15.3% | 10.7 | 7.1% | 4.6 | 10.8% | 8.8 |
| Change 2009–2018* | −2.5% | 4.6 | −3.9% | 3.6 | −3.3% | 4.1 |
| Awareness of useful reading strategies | | | | | | |
| Change 2009–2018 | −.015 | .124 | .089 | .094 | .042 | .119 |
| Percentage of students chatting daily | | | | | | |
| Prevalence in 2009 | 18.4% | 12.6 | 44.7% | 8.3 | 32.9% | 16.8 |
| Change 2009–2018 | 53.8% | 13.8 | 26.1% | 8.3 | 38.5% | 17.8 |

**The difference between both groups is not statistically significant ($p$ = .227).*
*Countries are weighed by the number of students they represent.*
*All other differences are statistically significant ($p$ < .05; two-tailed).*

**Table 1.**
*Reading literacy, reading fiction, reading strategies, and online chatting.*

survey findings fail to provide support for these findings. Data from the international PISA surveys indicate even slightly higher scores on the reading tests for 15-year-olds that frequently engage in online chatting [7]. Still, the notion that reading from screen hampers comprehension is supported by the finding that fewer students provide correct responses when test items are presented on computer screens instead in a paper format [8–10].

Unfortunately, the PISA surveys do not track *individual* reading literacy development over time. In each of the seven triennial surveys conducted during 2000–2018, the reading literacy of 15-year-olds in a particular year has been assessed for dozens of countries. Therefore, it is possible to map *per-country changes* in reading literacy. With regard to individual students, however, the surveys only yield cross-sectional data. Perhaps an effect of online chatting would be discerned if individual growth in reading literacy were assessed. Undoubtedly, there is an urgent need for research into the long-term effects of digital media on reading literacy [2].

## 4. The present study

The focus here is on per-country growth rates in online chatting and reading literacy trends. Their correlation points to an effect of changing environments. Superficial reading strategies may become the default for all individuals in a shared environment as online chatting spreads. In that case, everyone is affected to a similar extent. Such an effect would be similar to the effect of diesel and gasoline cars on health. Within a certain area, everyone is affected by exhaust gases. Individual variation in car use is hardly relevant.

The findings to be reported relate to changes during 2000–2018 and involve 39 countries from five continents. See **Figure 1** for an overview. **Table A1** provides a detailed list of the countries included. Special attention is paid to distinct patterns before and after 2009. The 2009 PISA survey was the first after the introduction of the iPhone in 2007. This innovation stimulated the exchange of information over the internet even further, in this case via compact hand-held devices. Moreover, in 2009 WhatsApp was launched. In a short time, this became the most popular online chatting tool in many countries across the world. In this study two groups of countries are compared:

1. Countries with high chatting prevalence in 2009 (more than 35% of 15-year-olds chatting daily) and decelerating growth afterward.

2. Countries with low chatting prevalence in 2009 (less than 35% of 15-year-olds chatting daily) and accelerating growth afterward.

The first group includes 29 countries and represents 55.2% of all students across the entire set of 39 countries. The second group consists of 10 countries representing 44.8% of all students. **Figure 1** shows that the group of countries with low chatting prevalence in 2009 mainly involves countries in East Asia, Southeast Asia, and Latin America. European and North American countries dominate the group with high chatting prevalence in 2009.

Both groups will be compared with regard to reading literacy trends during 2000–2009 and 2009–2018. In addition, correlations between per-country reading literacy trends and (changes) in chatting prevalence are reported. This study also aims to assess to what extent these correlations are mediated by changes in the frequency

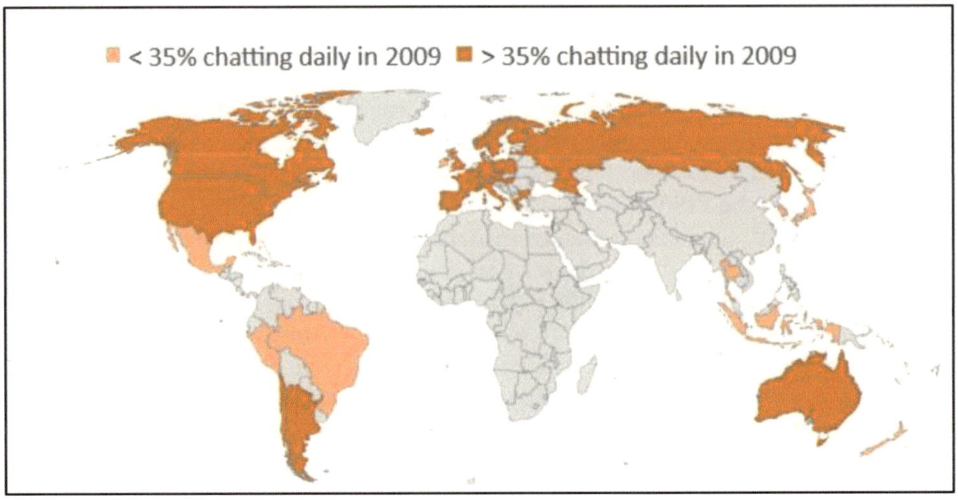


**Figure 1.**
*Countries included by chatting prevalence in 2009.*

of reading fiction and awareness of useful reading strategies. It seems plausible that increases in online chatting coincide with decreases in the reading of fiction. Online chatting involves swift exchanges of short and informal messages, whereas reading fiction requires (much) more time, effort, and patience. Given that online chat messages are usually (very) short, it also seems likely that increased online chatting promotes superficial reading and hampers the development of time-consuming reading strategies that allow for deeper comprehension. Prior research does report higher levels of reading comprehension among students that engage frequently in reading fiction [11]. Likewise, individuals that use appropriate reading strategies perform better on reading tests [12]. However, the novelty of this study, however, is that it shows to what extent *per-country* changes in these variables coincide with changes in reading literacy over time. The correlations' robustness is established through several additional data analyses controlling for potentially confounding variables.

### 4.1 The international PISA surveys

The Programme for International Student Assessment (PISA) is an Organisation for Economic Co-operation and Development (OECD) initiative. During 2000–2018 seven surveys have been conducted every 3 years. Due to the COVID pandemic, the survey planned for 2021 was postponed to 2022. Its primary objective is to assess the knowledge and skills of 15-year-olds in reading, mathematics, and science. In addition to cognitive tests, the data collection involves student questionnaires on school experiences and general background. PISA started in 2000 with 32 countries. Ten additional countries joined in 2002 [13].

Reading literacy was the main focus in 2000/2002, 2009, and 2018. In other years, reading literacy was assessed as well. However, most test items involved either mathematics or science in these surveys. Only when the surveys focused on reading, the student questionnaires addressed reading attitudes and reading-related activities in addition to general background questions.

The PISA data collection is highly standardized and therefore well-suited for cross-national comparisons. Across countries, students in the same age range are targeted. Age rather than grade level determines whether students are included in the national samples.

The present study involves 39 countries, which all took part in the first PISA survey. Two countries that participated in PISA 2000 were excluded. The participation in PISA of Liechtenstein ended in 2012. The Netherlands failed to meet all sample requirements in 2000. Romania, which joined in 2002, was excluded because the age range of the sampled students was not in line with that of the other countries.

To account for the variation in size among countries weights are applied. The samples of countries like Brazil, Indonesia, and the United States represent over two million students per year. On the other hand, for countries like Iceland and Luxembourg, this number is approximately five thousand. Country weights are based on the total number of students the national samples represent, averaged across years. The weights are standardized in such a way that their average equals one. This precludes that the statistical significance of the findings is overestimated. In the absence of weights, the European countries would dominate the results although they account for only 31% of all students represented by the national samples. The study includes 24 European countries. This would amount to 61% according to a one-country one-vote principle. For more details, see **Table A1**.

### 4.2 Measurements

The cognitive tests in PISA take up approximately 2 hours. The reading test addresses four main reading processes (reading fluently, locating information, understanding, and evaluating/reflecting) and includes several types of texts (e.g., narrations, argumentations, instructions). Test development is based on a conceptual framework that has been adapted a few times over the years [14]. Attention to digital environments and multiple resources was new to the 2018 reading assessment. An additional adaptation involves the switch from paper-and- pencil tests to computer-based ones in 2015. As a result, adjustments for "mode effects" were needed to take into account that computer-based test versions usually yield fewer correct responses. Notwithstanding these changes, assessment of trends in reading literacy over time is possible [15]. The average reading score across OECD countries is approximately 500 with a standard deviation around 100 points.

For this study, country-specific reading literacy trends were estimated for 2000–2009 and 2009–2018. These linear trends are based on each country's reading literacy averages in 2000, 2002, 2003, 2006, 2009, 2012, 2015, and 2018. A detailed account of reading trends, online chatting, and weights per country is provided in **Table A2**.

The main explanatory variable in the data analysis is per-country online chatting prevalence. In 2009 and 2018, students indicated how often they engaged in online chatting. The percentages of 15-year-olds that reported chatting several times a day are used to gauge changes in online chatting. Information on online chatting is lacking for 2000/2002. The 2009 percentages are considered to approximate growth during 2000–2009. It seems plausible to assume that at the turn of the century, the percentage of 15-year-olds chatting several a day was close to zero in each and every country. Chat platforms were around at the time, but it required much more effort and patience to get access to the internet than it does today. Connections were mainly made on desktop computers, via modems and analog telephone lines.

The student surveys in 2000/2002, 2009, and 2018 contained several items on reading activities. Using the answers related to reading fiction (including stories, novels, and narratives), per-country changes during 2000–2009 and 2009–2018 were calculated. These relate to the percentages of pupils who indicated that they read fiction at least once a month on a voluntary basis. For the countries that joined the first PISA survey in 2002, the difference between 2002 and 2009 was increased by $^9/_7$ in order to estimate change during 2000–2009.

Only the 2009 and 2018 questionnaires included items on reading strategies. As a result, changes in this variable can only be determined for the 2009–2018 period. Students were asked to rate the usefulness of six different strategies for understanding and memorizing a text (e.g., I underline important parts of the text). Next, it was established to what extent the student ratings concurred with a ranking established by reading experts [16]. Finally, this resulted in a score with zero mean and a standard deviation equal to one in OECD countries. High scores indicate student ratings that are in line with the experts' ranking.

## 5. Findings

### 5.1 Comparison between the two groups of countries

**Table 1** shows, by country group, the changes in reading literacy, percentages of students who read fiction at least once a month, awareness of useful reading strategies, and percentages of students who chat daily, both from 2000 to 2009 and 2009–2018. The bottom rows on online chatting show that the growth in online chatting strongly accelerated during 2009–2018 for the group of countries with low chatting prevalence in 2009. In contrast, the growth decelerated during 2009–2018 for the group with high chatting prevalence in 2009.

The countries with low chatting prevalence in 2009 showed substantial improvement in reading literacy during 2000–2009. The countries with high chatting prevalence in 2009 showed no improvement at all in the same period. The countries with low chatting prevalence in 2009 witnessed a strong acceleration in online chatting during 2009–2018. At the same time, reading literacy dropped. After a 21.2-point improvement during 2000–2009, reading literacy scores dropped by 9.9 points in the subsequent period. Online chatting growth decelerated during 2009–2018 in the countries with high chatting prevalence in 2009. Along with this deceleration, reading literacy improved by 9.2 points.

Percentages of students reading fiction at least once a month were on the rise in both groups of countries during 2000–2009. But, the increase was over twice as large in countries with low chatting prevalence in 2009 (15.3% points vs. 7.1%). The 2009–2018 period shows a modest decrease in reading fiction for both country groups. However, the difference between both groups is not statistically significant.

The countries with low chatting prevalence in 2009—and accelerated growth in the subsequent period—show a slight decline in awareness of useful reading strategies, whereas the countries with decelerating growth in online chatting during 2009–2018 show a clear improvement in awareness of useful reading strategies in that period.

### 5.2 Correlations between changes in the key variables

**Table 2** shows the correlations (Pearson's r) between changes in the key variables. The correlations below the diagonal involve changes during 2000–2009. The ones above the diagonal represent correlations between changes during 2009–2018. **Figures 2** and **3** provide a graphical display of the correlations between online

| | Percentage of students chatting daily | Reading literacy | Percentage of students reading fiction | Awareness of useful reading strategies |
|---|---|---|---|---|
| Percentage of students chatting daily | — | −.660 | .052* | −.552 |
| Reading literacy | −.574 | — | .057* | .726 |
| Percentage of students reading fiction | −.770 | .588 | — | −.098* |

**None of the correlations that involve reading fiction are significant for p < .05.*
*Countries are weighed by the number of students they represent.*
*Correlations for 2000–2009 are shown below the diagonal.*
*Correlations for 2009–2018 are shown above the diagonal.*

**Table 2.**
*Correlations (Pearson's r) between changes during 2000–2009 (below the diagonal) and 2009–2018 (above).*

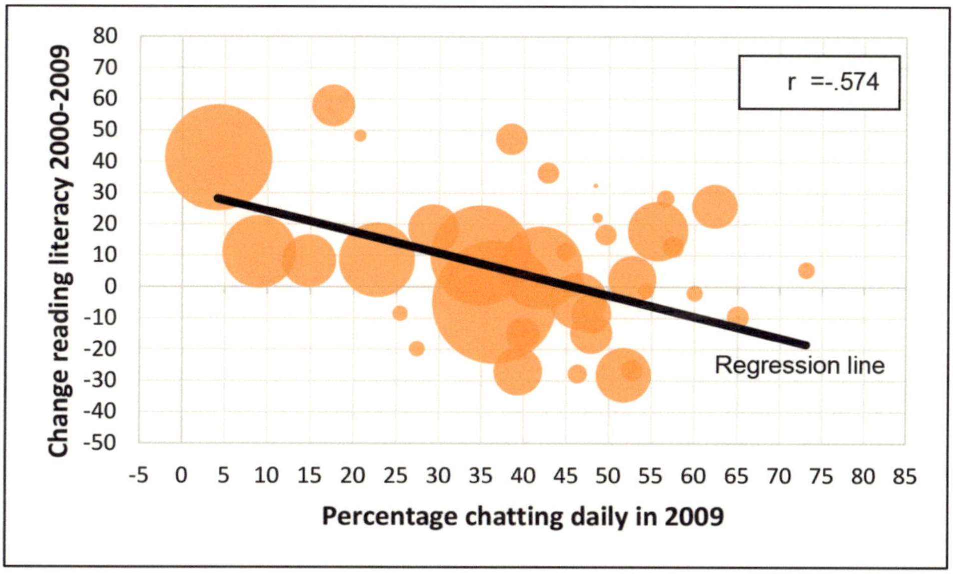


**Figure 2.**
*Online chatting prevalence 2009 and per-country reading change 2000–2009. The number of students per country determines the size of each bubble.*

chatting and changes in reading literacy during 2000–2009 and 2009–2018. Both graphs also show the regression lines that indicate to what extent reading literacy declines as online chatting increases. The following equations describe the bivariate linear regression equations for 2000–2009 and 2009–2018:

$$2000-2009: \text{Change reading literacy} = 31.038 - .676 \times \text{Chatting prevalence} \quad (1)$$

$$2009-2018: \text{Change reading literacy} = 27.633 - .700 \times \text{Change in chatting} \quad (2)$$

The above equations show the unstandardized regression coefficients. In a bivariate regression analysis, the standardized regression coefficient (beta) is equal to

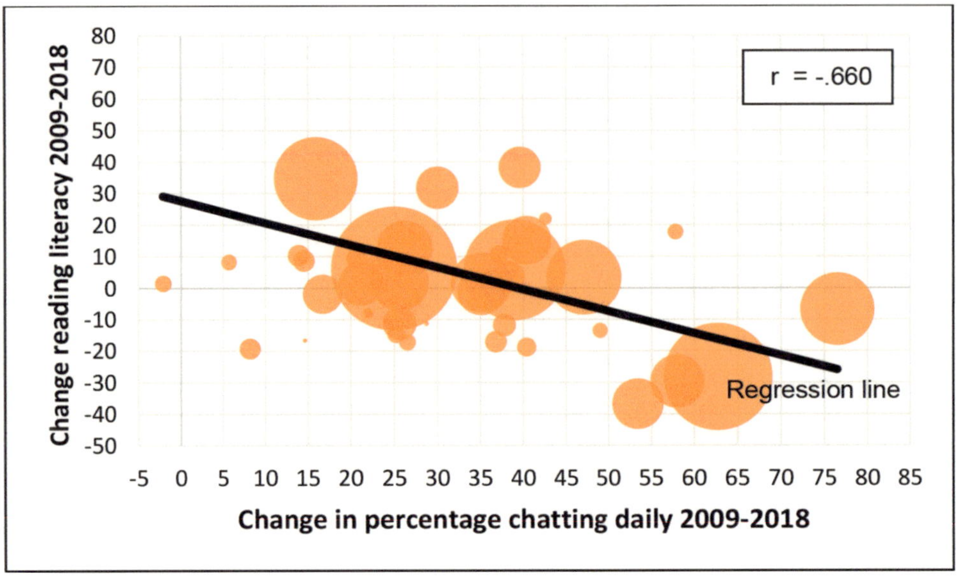


**Figure 3.**
*Change chatting 2009–2018 and per-country reading change 2009–2018. The number of students per country determines the size of each bubble.*

the correlation between the explanatory variable and the outcome variable. These correlations are reported in **Table 2**.

The unstandardized coefficients indicate that reading literacy drops by about seven-tenths of a point at a 1% increase in online chatting. The constants (intercepts) in the equations denote the expected reading literacy increase in the absence of any change in online chatting. Both intercepts are close to 30. This suggests that without the rise of online chatting, reading literacy would have increased by about 30 points in both periods. It can also be inferred from the equations that a 46% chatting prevalence in 2009 would coincide with zero reading change during 2000–2009. Belgium fits this scenario closely. By 2009, 45.8% of the Belgian 15-year-olds were engaged in online chatting several times a day. The change in reading literacy between 2000 and 2009 for Belgium is minimal (0.7 points, see **Table A2**). During 2009–2018, a 39% increase in chatting would match a situation without any change in reading literacy. The findings for Chile are in line with this scenario. Reading literacy improved only marginally in Chile during 2009–2018 (3.6 points). At the same time, online chatting increased by 38.3 percent points.

The figures in **Table 2** confirm that in both periods the correlation between (changes in) the percentage of students chatting online and reading literacy changes is strong and negative (−.574; −.660). The correlations suggest that during 2000–2009 change in reading fiction might present an intermediary factor in the relationship between online chatting and reading literacy. Change in reading fiction correlates strongly with both online chatting and reading literacy. Change in awareness of useful reading strategies seems to assume this role during 2009–2018.

### 5.3 The search for mediating and confounding variables

A number of regression analyses have been conducted to further explore the conjecture that reading fiction and reading strategies function as mediating factors.

If the relation between online chatting and reading literacy is (partially) mediated by reading fiction or reading strategies, multiple regression analyses with both online chatting and reading fiction/reading strategies as explanatory variables will show a considerably reduced coefficient for online chatting. The high correlations between online chatting and changes in reading fiction (during 2000–2009) or reading strategies (during 2009–2018) suggest strong relationships between online chatting and both variables. If multiple regression analysis fails to show strong coefficients for online chatting, the findings may still be in line with an indirect effect. Online chatting is clearly related to decreases in percentages of students reading fiction and awareness of useful reading strategies. In turn, these variables may show strong coefficients in a multiple regression analysis.

In the next step, it is investigated to what extent 15 possibly confounding variables can account for the correlations in **Table 2**. They include societal changes (e.g., regarding parents' education levels, internet availability), educational changes (e.g., student-teacher ratios), changes in sample characteristics (e.g., coverage of all 15-year-olds), and additional changes in reading activities and attitudes (e.g., enjoyment of reading). See **Table A3** for a complete list of the control variables. By means of stepwise regression, it is established to what extent they are related to changes in reading literacy, reading fiction, and awareness of reading strategies. With regard to reading fiction and reading strategies, changes in reading attitudes and reading activities were not included as control variables. These variables can be either effects or causes of reading fiction and awareness of useful reading strategies.

**Table 3** shows the results of the first set of multiple regression analyses. For the 2000–2009 period, the findings show that the coefficient of online chatting prevalence in 2009 with change in reading literacy is no longer statistically significant when change in reading fiction is included as well. The p-value for the regression coefficient of reading fiction can still be considered significant at the .05 level in a one-tailed test. This conclusion is warranted as the coefficient indicates a positive relationship, which is in line with expectations. Regarding the 2009–2018 period, both the regression coefficients of change in online chatting and awareness of useful reading strategies reach statistical significance. However, also in this case the standardized regression coefficient of online chatting is considerably smaller than the bivariate correlation in **Table 2** (−.373 vs. -.660).

**Table 4** shows the results of the stepwise regression for the 2000–2009 period. With regard to change in reading fiction, stepwise regression fails to produce any

| | **beta** | **p-value** | **$R^2$** |
|---|---|---|---|
| Outcome: Change reading literacy 2000–2009 | | | |
| Prevalence online chatting 2009 | −.297 | .156 | |
| Change reading fiction 2000–2009 | .359 | .089 | .382 |
| Outcome: Change reading literacy 2009–2018 | | | |
| Change online chatting 2009–2018 | −.373 | .004 | |
| Change reading strategies 2009–2018 | .520 | <.001 | .624 |

*Countries are weighed by the number of students they represent.*
*p-values are based on two-tailed tests.*

**Table 3.**
*Multiple regression analysis—Standardized coefficients (beta) of online chatting and reading fiction/reading strategies.*

| | beta | p-value | $R^2$ |
|---|---|---|---|
| Outcome: Change reading literacy 2000–2009 | | | |
| Prevalence online chatting 2009 | — | — | |
| Change reading fiction 2000–2009 | .414 | <.001 | |
| Change student-teacher ratio 2000–2009 | –.413 | <.001 | |
| Change coverage all 15-year-olds 2000–2009 | .322 | .003 | |
| Change frequency reading newspapers 2000–2009 | .228 | .028 | .751 |
| Outcome: Change reading fiction 2000–2009 | | | |
| Prevalence online chatting 2009 | –.770 | <.001 | .594 |

*Countries are weighed by the number of students they represent.*
*p-values are based on two-tailed tests.*

**Table 4.**
*Stepwise regression analysis regarding changes during 2000–2009.*

significant coefficients in addition to the prevalence of online chatting in 2009. None of the possibly confounding factors that were considered can account for changes in the percentages of students reading fiction at least once a month.

For change in reading literacy, the stepwise regression first of all yields no significant coefficient for online chatting. This implies that when controlling for possibly confounding factors, no direct effect on online chatting can be discerned. On the other hand, the results do show a significant coefficient for reading fiction. This coefficient is smaller than the bivariate correlation (.414 vs. .588), but still substantial and statistically significant. Additional variables with a significant coefficient are changes in student-teacher ratios, coverage of all 15-year-olds, and frequency of reading newspapers.

All in all, the findings indicate a strong and negative effect of online chatting on changes in reading fiction, which in turn shows a substantial effect on changes in reading literacy during 2000–2009. In other words, the negative relationship between increased online chatting and reading literacy runs via decreased percentages of students reading fiction during 2000–2009.

The findings for 2009–2018 are reported in **Table 5**. With regard to change in reading strategies, the stepwise regression yields one additional variable with a significant coefficient, namely change in coverage of all 15-year-olds in the national samples. Awareness of useful reading strategies decreased in countries with increasing coverage during 2009–2018. However, the analysis also yields a significant and negative coefficient for change in online chatting. It differs only slightly from the bivariate correlation between online chatting and reading literacy (–.533 vs. -.552).

Regarding changes in reading literacy during 2009–2018, the stepwise regression again does not yield a significant effect of online chatting. But, the analysis does show a strong coefficient for changes in reading strategies. It is not as strong as the bivariate correlation (.584 vs. .726), but definitely substantial. Additional variables showing significant coefficients are changes in parents' occupation and reading enjoyment.

Also for the 2009–2018 period, the findings are in line with the notion that the negative relationship between online chatting and changes in reading literacy is indirect. Increases in online chatting coincide with decreasing awareness of useful reading strategies. Subsequently, changing awareness of useful reading strategies coincide with changes in reading literacy.

| | beta | p-value | $R^2$ |
|---|---|---|---|
| Outcome: Change reading literacy 2009–2018 | | | |
| Change online chatting 2009–2018 | — | — | |
| Change reading strategies 2009–2018 | .584 | <.001 | |
| Change parents' occupation 2009–2018 | .372 | <.001 | |
| Change reading enjoyment 2009–2018 | .281 | .005 | .700 |
| Outcome: Change reading strategies 2009–2018 | | | |
| Change online chatting 2009–2018 | −.516 | <.001 | |
| Change coverage all 15-year-olds 2009–2018 | −.553 | <.001 | .407 |

*Countries are weighed by the number of students they represent.*
*p-values are based on two-tailed tests.*

**Table 5.**
*Stepwise regression analysis regarding changes during 2009–2018.*

## 6. Conclusion and discussion

The findings show that developments in reading literacy levels of 15-year-olds across the world in the first two decades of the twenty-first century are clearly linked to the speed with which online chatting has spread. Reading literacy markedly improved during 2000–2009 in countries with low percentages of 15-year-olds chatting on a daily basis. In the subsequent period, the prevalence of online chatting increased rapidly in these countries. This coincided with a clear decrease in reading literacy. In contrast, countries with high chatting prevalence in 2009 showed no improvement during 2000–2009. When the growth of online chatting clearly decelerated in these countries during 2009–2018, reading literacy improved. The relationship between online chatting and reading literacy was found to be mediated by changes in reading fiction during 2000–2009 and by changes in awareness of useful reading strategies during 2009–2018. The finding that these factors mediate the relationship between online chatting and reading literacy is clearly in line with theoretical considerations on the relationship between digital media use and reading literacy.

By 2009, online chatting had already become popular among 15-year-olds, especially in Europe and North America. In East Asia, Southeast Asia, and Latin America, the percentages of students chatting on a daily basis were still fairly moderate in 2009. It seems that the launch of the iPhone in 2007 and WhatsApp in 2009 triggered the rise of online chatting in those parts of the world.

The relationship between online chatting and reading literacy suggests a large effect. During 2000–2009, countries with low chatting prevalence showed an improvement of 21.2 points. Countries with high prevalence experienced a minute decrease (−1.2, see **Table 1**). This amounts to a difference of over 20 points on the PISA reading literacy scale. In the subsequent period, the difference between both groups amounts to 19 points (−9.9 vs. 9.2). For both periods, this implies a difference of about two-tenth standard deviation in the distribution of individual-level reading scores. In the field of education, this may be considered a large effect [17].

The data analysis also indicates that a moderate increase in online chatting does not necessarily pose a threat to reading literacy. An increase below 40 percentage points over a nine-year period was found to coincide with zero reading literacy change. This implies that at a limited increase per year (4–5%) reading literacy may remain stable.

In most countries included in the present study, the percentage of students chatting several times a day already exceeded 60% in 2018. The Russian Federation and Peru are the only exceptions. This suggests that there may be not much room left for further increases. Per-country percentages of students chatting several times a day have proven to be a useful indicator to gauge the impact of digital media use on reading literacy in the first two decades of the present century. However, it seems likely that future research will need alternative and more fine-grained measures.

The present study involves correlations between changes in country aggregates over time. Therefore, the conclusions relate exclusively to the country level. The findings do *not* show that individuals that frequently engage in online chatting are poor readers. This study only shows that per-country increases in online chatting in the first two decades of the twenty-first century coincided with declining reading literacy trends. At the country level, the PISA data allow for an analysis of changes over time, but this is not possible for the individual-level PISA data. The correlations reported in this chapter point to the effects of changing environments, like the effect of car use on the prevalence of lung-related health problems. However, the present study stands out as it shows that per-country changes in online chatting, frequency of reading fiction, and awareness of useful reading strategies coincide with changes in reading literacy. The findings are based on correlations. As a result, it cannot be precluded that other variables that were not controlled for in data analysis could account for the correlations that were reported. The variables that were controlled for showed some additional correlations with reading literacy, frequency of reading fiction, and/or awareness of reading strategies. However, this did not lead to substantially different conclusions. In addition, it also possible that country-specific circumstances might account for changes in reading literacy, e.g., major curriculum changes or local teacher shortages.

The research findings indicate how the diffusion of online chatting has shaped per-country reading literacy developments. This faces the education sector with new challenges. In comparison to a few decades ago, today's youth acquire fewer reading skills outside school. Even more time and effort will have to be put into reading education than before by schools and teachers to prevent further decline in traditional reading skills. Still, the impact of online chatting may turn out trivial compared to recent developments in the field of Artificial Intelligence (AI). Currently, chatbots are improving rapidly and can generate texts that are barely distinguishable for human readers from texts written by fellow humans. In addition, they are very well capable of responding to prompts and questions that are posed in natural language. AI is getting better and better both at generating and interpreting natural language. This raises the question of which reading and writing skills people in the future will still need. It seems conceivable that in the future exchange of information and ideas will largely take place via ICT tools developed through machine learning (i.e., computer algorithms that are fine-tuned to the extreme by computer algorithms). Human input may become more and more redundant.

It goes without saying that the spread of ICT has had an enormous impact on people's lives all over the world. In many respects, life has become easier and more enjoyable thanks to countless innovative ICT applications. Still, every major transition comes with certain drawbacks. Loss of traditional reading literacy skills seems to be one of them. It would be unfortunate if thorough and traditional reading skills were lost. For the moment, old-school reading comprehension skills still prove to be very useful when searching and interpreting online information [18, 19].

## Appendix

| | Low chatting prevalence (< 35%) in 2009 | High chatting prevalence (> 35%) in 2009 |
|---|---|---|
| Africa | | |
| No countries included | — | — |
| Asia | | |
| 6 countries | 4 countries | 2 countries |
| 25.4% of all students | 24.6% of all students | 0.8% of all students |
| | Indonesia, Japan, South Korea, Thailand. | Hong Kong, Israel. |
| Australasia | | |
| 2 countries | 1 country | 1 country |
| 1.5% of all students | 0.3% of all students | 1.2% of all students |
| | New Zealand | Australia |
| Europe | | |
| 24 countries | 2 countries | 22 countries |
| 31.1% of all students | 0.4% of all students | 30.6% of all students |
| | Albania, Ireland | Austria, Belgium, Bulgaria, Czech Republic, Denmark, Finland, France, Germany, Greece, Hungary, Iceland, Italy, Latvia, Luxembourg, Norway, Poland, Portugal, Russian Federation, Spain, Switzerland, Sweden, United Kingdom |
| North America | | |
| 3 countries | 1 country | 2 countries |
| 25.0% of all students | 6.2% of all students | 18.8% of all students: |
| | Mexico | Canada, United States |
| South America | | |
| 4 countries | 2 countries | 2 countries |
| 17.0% of all students | 13.3% of all students | 3.7% of all students |
| | Brazil, Peru | Argentina, Chile |
| Total | 10 countries | 29 countries |
| | 44.8% of all students | 55.2% of all students |

**Table A1.**
*Countries with low vs. high chatting prevalence by continent.*

| | Reading literacy | | Online chatting | | Country Weight |
|---|---|---|---|---|---|
| | Change 2000–2009 | Change 2009–2018 | Percentage 2009 | Change 009–2018 | |
| Asia | | | | | |
| Hong Kong | 28.3 | −14.7 | 56.6 | 25.2 | .132 |
| Indonesia | 41.2 | −28.2 | 4.2 | 62.7 | 4.956 |
| Israel | 36.5 | 5.4 | 42.8 | 38.1 | .199 |
| Japan | 11.2 | −6.8 | 9.0 | 76.5 | 2.302 |
| South Korea | 18.4 | −36.8 | 29.3 | 53.4 | 1.100 |
| Thailand | 8.3 | −29.7 | 14.8 | 58.0 | 1.230 |
| Australasia | | | | | |
| Australia | −15.1 | −11.3 | 39.9 | 25.6 | .475 |
| New Zealand | −8.4 | −13.6 | 25.4 | 49.0 | .102 |
| Europe | | | | | |
| Albania | 48.3 | 21.9 | 20.7 | 42.6 | .066 |
| Austria | −27.8 | 10.6 | 46.3 | 37.2 | .158 |
| Belgium | 0.7 | −11.8 | 45.8 | 37.8 | .227 |
| Bulgaria | 5.4 | 1.5 | 73.1 | −2.1 | .122 |
| Czech Republic | −9.5 | 10.3 | 65.0 | 13.9 | .207 |
| Denmark | −1.1 | 7.4 | 54.3 | 22.6 | .112 |
| Finland | −11.9 | −17.3 | 48.5 | 26.5 | .116 |
| France | −4.4 | 1.8 | 46.4 | 25.6 | 1.414 |
| Germany | 17.8 | 1.7 | 55.7 | 35.1 | 1.565 |
| Greece | 8.1 | −17.1 | 38.4 | 36.8 | .202 |
| Hungary | 12.8 | −19.3 | 57.5 | 8.2 | .192 |
| Iceland | −9.3 | −16.7 | 65.0 | 14.6 | .008 |
| Ireland | −19.8 | 17.8 | 27.4 | 57.8 | .109 |
| Italy | 2.3 | −1.1 | 52.7 | 35.1 | .991 |
| Latvia | 22.1 | −8.0 | 48.6 | 22.0 | .046 |
| Luxembourg | 32.5 | −11.3 | 48.4 | 28.7 | .010 |
| Norway | −2.0 | 8.2 | 60.0 | 5.7 | .109 |
| Poland | 25.8 | 1.4 | 62.3 | 21.2 | .860 |
| Portugal | 16.7 | 9.4 | 49.6 | 20.7 | .188 |
| Russian Federation | 6.1 | 34.7 | 42.0 | 15.9 | 3.003 |
| Spain | −14.7 | 38.2 | 47.9 | 39.6 | .753 |
| Sweden | −26.6 | 8.6 | 52.7 | 14.5 | .201 |
| Switzerland | 11.2 | −18.8 | 44.9 | 40.4 | .157 |
| United Kingdom | −28.2 | 12.7 | 51.7 | 26.2 | 1.302 |

| | **Reading literacy** | | **Online chatting** | | **Country Weight** |
|---|---|---|---|---|---|
| | **Change 2000–2009** | **Change 2009–2018** | **Percentage 2009** | **Change 009–2018** | |
| North America | | | | | |
| Canada | −8.6 | −2.0 | 47.9 | 16.7 | .676 |
| Mexico | 8.6 | 3.3 | 22.7 | 47.1 | 2.432 |
| United States | −4.9 | 6.2 | 36.5 | 25.0 | 6.645 |
| South America | | | | | |
| Argentina | −27.0 | 15.1 | 39.3 | 40.4 | 1.022 |
| Brazil | 10.0 | 5.7 | 34.9 | 39.0 | 4.404 |
| Chile | 47.2 | 3.6 | 38.5 | 38.3 | .437 |
| Peru | 57.9 | 31.8 | 17.6 | 30.0 | .772 |

**Table A2.**
*Key data per country.*

| **Societal developments** | **Educational changes** | **Changes in sample characteristics** | **Changes in reading behavior & attitudes** |
|---|---|---|---|
| Percentage immigrants | Student-teacher ratio | Percentage female students | Reading enjoyment |
| Parents' education | Grade level of 15-year-olds | Coverage of all 15-year-olds | Frequency reading magazines |
| Parents' occupation | | Student age | Frequency reading comic books |
| Internet availability | | | Frequency reading non-fiction books |
| | | | Frequency reading newspapers |
| | | | Time reading for enjoyment |

**Table A3.**
*Control variables in stepwise regression analyses.*

## Author details

Hans Luyten
University of Twente, Enschede, Netherlands

*Address all correspondence to: j.w.luyten@utwente.nl

## References

[1] Carr NG. The Shallows: What the Internet Is Doing to our Brains. W. W Norton; 2010

[2] Wolf M. Reader, Come Home: The Reading Brain in a Digital World. New York: Harper Collins; 2018

[3] OECD. 21st-Century Readers: Developing Literacy Skills in a Digital World. Paris: OECD Publishing; 2021. DOI: 10.1787/a83d84cb-en

[4] Clinton V. Reading from paper compared to screens: A systematic review and meta-analysis. Journal of Research in Reading. 2019;**42**(2):288-325. DOI: 10.1111/1467-9817.12269

[5] Delgado P, Vargas C, Ackerman R, Salmerón L. Don't throw away your printed books: A meta-analysis on the effects of reading media on reading comprehension. Educational Research Review. 2018;**25**:23-38. DOI: 10.1016/j.edurev.2018.09.003

[6] Kong Y, Seo YS, Zhai L. Comparison of reading performance on screen and on paper: A meta-analysis. Computers & Education. 2018;**123**:138-149. DOI: 10.1016/j.compedu.2018.05.005

[7] Luyten H. The global rise of online chatting and its adverse effect on reading literacy. Studies in Educational Evaluation. 2022;**72**:1-7. DOI: 10.1016/j.stueduc.2021.101101

[8] Jerrim J, Micklewright J, Heine J-H, Sälzer C, McKeown C. PISA 2015: How big is the 'mode effect' and what has been done about it? Oxford Review of Education. 2018;**44**:476-493. DOI: 10.1080/03054985.2018.1430025

[9] OECD. PISA 2015 Technical Report. OECD Publishing; 2017 Available from: https://www.oecd.org/pisa/data/2015-technical-report/PISA2015_TechRep_Final.pdf

[10] Robitzsch A, Lüdtke O, Goldhammer F, Kroehne U, Köller O. Reanalysis of the German PISA data: A comparison of different approaches for trend estimation with a particular emphasis on mode effects. Frontiers in Psychology. 2020;**11**:884. DOI: 10.3389/fpsyg.2020.00884

[11] Jerrim J, Moss G. The link between fiction and teenagers' reading skills: International evidence from the OECD PISA study. British Educational Research Journal. 2019;**45**(1):181-200. DOI: 10.1002/berj.3498

[12] Artelt C, Schiefele U, Schneider W. Predictors of reading literacy. European Journal of Psychology of Education. 2001;**16**:363-383. DOI: 10.1007/BF03173188

[13] OECD. Knowledge and Skills for Life: First Results from PISA 2000. Paris: OECD Publishing; 2001. DOI: 10.1787/9789264195905-en

[14] OECD. PISA 2018 Assessment and Analytical Framework. Paris: OECD Publishing; 2019. DOI: 10.1787/b25efab8-en

[15] Schleicher A. PISA 2018, Insight and Interpretations. Paris: OECD Publishing; 2019. Available from: https://apo.org.au/node/270241

[16] OECD. PISA 2009 Technical Report. Paris: OECD Publishing; 2012. DOI: 10.1787/9789264167872-en

[17] Kraft MA. Interpreting Effect Sizes of Education Interventions

(Ed Working Paper #19-10). Providence, RI: Annenberg Institute at Brown University; 2019. Available from: https://files.eric.ed.gov/fulltext/ED602384.pdf

[18] Kanniainen L, Kiili C, Tolvanen A, Aro M, Leppänen PHT. Literacy skills and online research and comprehension: Struggling readers face difficulties online. Reading and Writing. 2019;**32**:2201-2222. DOI: 10.1007/s11145-019-09944-9

[19] Salmerón L, García A, Vidal-Abarca E. The development of adolescents' comprehension-based internet reading activities. Learning and Individual Differences. 2018;**61**:31-39. DOI: 10.1016/j.lindif.2017.11.006

Chapter 8

# Mines and Machines: Virtual Assistants in Modern Mining

*Ali Soofastaei*

## Abstract

"Mines & Machines: Virtual Assistants in Modern Mining" delves into the transformative power of technology, specifically focusing on the role of virtual assistants in revolutionizing the mining industry. This chapter looks at how Artificial Intelligence, Machine Learning, and virtual assistants alter the traditional landscapes of mining operations. From improving safety protocols to optimizing resource extraction, these technologies are at the forefront of innovation, efficiency, and sustainability. The chapter begins by providing historical context, underscoring the mining industry's rich tapestry of evolution and its enduring significance in human civilization. It then transitions into the modern era, explaining today's mining industry's challenges, including environmental considerations and ethical concerns. Central to this discourse is exploring how virtual assistants are poised to address these challenges. By presenting case studies, empirical data, and expert interviews, the chapter's content makes a compelling argument for the positive impacts of implementing virtual assistants in mining operations. Topics include real-time data analysis, predictive maintenance, safety monitoring, and environmental conservation. Aimed at industry professionals, researchers, and policy-makers, "Mines & Machines" serves as both a resourceful guide and a catalyst for discussing the future of mining in the age of artificial intelligence. It invites the reader to consider the broader implications of these technologies, not only for the mining industry but also for society at large.

**Keywords:** mining industry, virtual assistants, modern mining, mining 4.0, artificial intelligence, advanced analytics, information technology, prediction, optimization, making intelligent decisions, productivity, improvement

## 1. Introduction

As an industry, mining boasts one of the wealthiest tapestries of history among all human endeavors, a story that unfolds across millennia and continents. Not merely content with carving tunnels deep into the Earth's crust, mining has played a memorable role in shaping the trajectory of civilizations and their geopolitical landscapes. From the earliest human societies that extracted flint to craft rudimentary tools to the rise of advanced civilizations that depended on metallurgy for weapons and artifacts, mining has been a consistent thread weaving through the fabric of human progress.

Moreover, mining has catalyzed technological innovation and scientific discovery. Think of the steam engine, originally developed to pump water out of mines, which later revolutionized transportation and manufacturing. Or consider the countless minerals and rare earth elements crucial for modern electronics and renewable energy technologies; these materials are brought to the surface through increasingly sophisticated mining techniques.

The industry has also had profound cultural, social, and economic impacts. Entire towns and cities have been birthed around mining operations, creating communities with unique cultures and traditions. Economically, extracting precious metals like gold and silver has often led to periods of prosperity, funding monumental architecture, art, and even military campaigns.

However, it's essential to acknowledge the environmental and ethical challenges that mining has posed, including ecological degradation and labor exploitation. The industry's history is complex and multifaceted, continually balancing its potential for immense benefit against its risks and consequences.

As we look to the future, mining continues to evolve, adopting more sustainable practices and leveraging cutting-edge technologies like automation and artificial intelligence to minimize its negative footprint. Still, its role as a cornerstone in the edifice of human civilization remains as relevant today as it was thousands of years ago [1].

### 1.1 The historical context of mining

Historically, the act of mining was both a symbol of human ambition and a testament to human endurance. The ancient Egyptians mined malachite and other minerals from the Sinai Peninsula as far back as 3000 BCE, with the treasures they extracted becoming the cornerstone of their great civilization. In the Americas, the vast silver mines of Potosí in present-day Bolivia funded the Spanish Empire's dreams and ambitions in the sixteenth century.

But beyond the wealth and the geopolitics, mining has always been about the people. Often facing difficult conditions, Miners have toiled in the depths for the promise of a better tomorrow. These subterranean endeavors have brought tales of heroism, tragedy, community, and resilience. Entire towns and cultures have risen around mines, each with unique traditions and stories.

However, like all industries spanning millennia, mining has faced its fair share of challenges. The balance between extracting value from the earth and maintaining its environmental and social sustainability has been a perennial concern [2].

### 1.2 The digital revolution in industry

Fast forward to the late twentieth and early twenty-first centuries, and another revolution was taking shape—the digital process. The world was rapidly moving from analog systems to digital interfaces, and industries were transforming. From manufacturing to healthcare, the digital age ushered in efficiencies previously deemed impossible.

Data became the new gold in this digital era, and algorithms were the new mining tools. Companies began to harness the power of technology, utilizing vast amounts of data to optimize processes, reduce costs, and enhance productivity. With its intricate operations and high stakes, the mining industry stood to gain immensely from this shift. The introduction of digital tools allowed for more precise exploration, better safety protocols, and overall more efficient operations [3].

Yet, this digital transition was more than just about efficiency. It signified a profound change in how industries viewed their place globally. The emphasis shifted towards sustainability, with technology providing tools to minimize environmental impact. Automation and virtual assistants began taking center stage, promising operational excellence and a more harmonious relationship with our planet.

This chapter delves into the fascinating intersection of these two worlds: the time-honored mining industry and the ground-breaking innovations of the digital age. We'll explore how virtual assistants, powered by advanced algorithms and cutting-edge technology, redefine the future of mining. As we embark on this journey, we'll discover the transformative potential of this synergy and, perhaps, catch a glimpse of the industry at large.

## 2. Mining in a modern world

As we traverse the annals of history and the promise of tomorrow, we must recognize the meeting points of past practices and future innovations. The most significant shifts often occur at these junctures, bridging traditions with modernity and legacy methods with cutting-edge solutions.

### 2.1 A global perspective

With globalization in full swing by the late twentieth century, mining operations were not just local or regional endeavors anymore; they became global enterprises. As multinational corporations sought resources beyond their borders, the scope and scale of mining operations expanded dramatically. This international shift brought with it both opportunities and challenges. On the one hand, there were new territories to explore, with vast untapped resources. Conversely, the complexities of working in diverse environments, cultures, and regulatory landscapes became apparent.

The digital revolution could not have come at a better time. It offered solutions to many of the challenges faced by the global mining industry. Real-time communication, data sharing across continents, satellite imagery for exploration, and automated machinery for remote operations became invaluable tools. Digital platforms enabled seamless collaboration between teams situated thousands of miles apart, ensuring that expertise could be shared and solutions could be co-created.

### 2.2 Societal impacts of the digital wave

Beyond the operational realm, the societal impact of the digital revolution on the mining industry was palpable. Communities around mining sites, once isolated, were now connected to the broader world. This connectivity empowered them with information and gave them a voice in the global conversation about sustainable and ethical mining practices.

Moreover, the influx of technology brought about new education and skill development in the far-flung mining regions, leading to overall societal development. Mining towns and communities saw the emergence of training centers focused on digital tools and automation. Once reliant solely on physical prowess and experience, Miners had to become adept at interfacing with technology, analyzing data, and making decisions based on digital insights.

## 2.3 The promise and the responsibility

The convergence of mining and digital technology, encapsulated by the rise of virtual assistants, promises more efficient, sustainable, and safer mining practices. But with this promise also comes a responsibility. As we lean into a future where machines play an increasing role in our industries, ensuring that the human element, with all its intricacies, experiences, and wisdom, is not left behind is imperative.

It's a journey of balance, of marrying the old with the new, and ensuring that as we dig deeper into our planet, we also elevate the communities and people that have long been the backbone of the mining industry.

As we move forward in this exploration, we invite readers to approach this convergence with an open mind, recognizing the vast potentials and the challenges and appreciating the intricate dance of tradition and innovation.

## 2.4 Embracing the digital frontier

In any industry's evolution, there comes a time when a paradigm shift redefines the very core of its operations. For mining, introducing digital technologies and virtual assistants represents such a moment. However, understanding this shift requires more than a cursory glance at new technologies. It requires deep diving into the industry's ethos and stakeholders' aspirations.

## 2.5 A new breed of miners

The miner of the twenty-first century is no longer just an individual wielding a pickaxe; they are technologists, environmentalists, data analysts, and community leaders. The modern mining environment has become a melting pot of interdisciplinary collaboration. Geologists use advanced sensors and satellite data to pinpoint mineral deposits, while data scientists employ artificial intelligence to predict equipment failures, optimize processes, and ensure safety. All these changes signify a shift in tools, methods, and mindset.

## 2.6 The environment and sustainability

As our world grapples with the pressing issues of climate change and environmental degradation, the mining sector is now at a crossroads. The digital revolution offers a path toward more sustainable practices. With the aid of virtual assistants, environmental monitoring becomes more precise. Real-time data allows for quick responses to any anomalies, minimizing harm. Moreover, digital simulations can predict the ecological outcome of certain mining activities, enabling companies to opt for the most eco-friendly approach.

## 2.7 Community engagement in the digital age

For centuries, mining operations have been closely intertwined with communities. With the onset of the digital age, this relationship is transforming. Virtual town halls, digital feedback systems, and community-driven data collection foster a more inclusive dialog between miners and residents. This collaborative approach ensures that the benefits of mining are equitably shared and the concerns of local inhabitants are addressed promptly.

### 2.8 The ethical imperatives

With great power comes great responsibility. The capabilities of digital technologies are immense, but they also present ethical dilemmas. Data privacy, job displacement due to automation, and the potential misuse of technology are genuine concerns the industry must navigate. The onus is on both the creators and users of these technologies to ensure they are employed responsibly and ethically.

### 2.9 A vision forward

As we delve deeper into the world of Mines and machines, our journey will not only be about understanding technological marvels but also about grasping the profound societal implications of this convergence. We're standing at the cusp of a new era—one that holds the potential to reshape the mining industry for the better. Together, through the pages of this exploration, we will uncover the challenges, celebrate the triumphs, and envision a future where technology and humanity coalesce to unearth treasures, both from the earth and within our collective potential.

## 3. Emergence of virtual assistants in mining

The transformative power of technology is nowhere more evident than in the rise and adaptation of virtual assistants. From answering simple questions on our smartphones to orchestrating complex operations in industries, virtual assistants have rapidly evolved to become indispensable assets in various fields. In the mining sector, their emergence marks a significant shift toward increased efficiency, safety, and sustainability [4].

### 3.1 Evolution of virtual assistants: From homes to heavy industry

The genesis of virtual assistants can be traced back to personal computing and the desire to make human-computer interaction more natural and intuitive. Early voice recognition software was clunky and limited but set the stage for what would come.

With the rise of smartphones, companies like Apple, Google, and Amazon introduced Siri, Google Assistant, and Alexa, respectively. These were designed for personal use—setting reminders, playing music, or searching the web using voice commands.

These consumer-oriented virtual assistants' success led to realizing their potential in specialized fields. Their adaptability meant they could be trained for specific industries, from healthcare to finance and heavy industries like mining.

Despite its ancient roots, mining has always been at the forefront of adopting technologies that boost productivity and safety. Virtual assistants entered this space, offering solutions like equipment monitoring, predictive maintenance, and data-driven decision-making. Their ability to quickly process vast amounts of data made them a natural fit for the complex and data-rich mining environment.

## 4. Core technologies powering virtual assistants

1. Artificial Intelligence (AI) and Machine Learning (ML):

At the heart of any virtual assistant is AI. In mining, ML algorithms can predict machinery failures by analyzing patterns in operational data, thus reducing downtime and increasing safety.

2. Natural Language Processing (NLP):

NLP allows virtual assistants to understand, interpret, and generate human language. This is crucial in mining environments where quick, clear communication can be a matter of utmost safety.

3. Internet of Things (IoT):

Countless sensors collect data in mining operations, from temperature readings in underground tunnels to vibration measurements on heavy machinery. Virtual assistants utilize IoT to constantly monitor these sensors, making real-time decisions and alerts based on the incoming data.

4. Cloud computing:

The vast data generated in mining operations must be stored, processed, and accessed efficiently. Cloud platforms provide the necessary infrastructure for virtual assistants to operate seamlessly, ensuring data is available whenever and wherever it's needed.

5. Augmented Reality (AR) and Virtual Reality (VR):

In mining, AR and VR are leveraged for training purposes and to assist in complex tasks. Virtual assistants can guide technicians through equipment repairs using AR overlays, ensuring procedures are followed accurately.

In conclusion, the emergence of virtual assistants in mining is a testament to technological advancement and indicates the industry's commitment to innovation, safety, and efficiency. As these technologies continue to evolve, so too will their role in shaping the future of mining. Fusing age-old practices with cutting-edge technology promises a brighter, safer, and more sustainable future for the mining industry.

## 5. The mining landscape

Mining stands as a cornerstone industry, the crucial link between the Earth's untapped natural resources and the relentless demands of modern civilization. This industry is vast and intricate, with its topography constantly evolving to mirror ongoing technological innovations and societal shifts. From its deep-rooted, time-honored traditions that trace back to the dawn of human civilization to the contemporary call for digital integration and automation, the mining landscape is a fascinating tableau of change and continuity.

The sector has been marked by a long and storied history, with each era contributing its tools, techniques, and philosophies. From the early days of manual labor and simple tools to extract valuable ores to the industrial revolution that mechanized many mining processes, each advancement significantly improved efficiency, productivity, and worker safety.

In today's context, the industry is experiencing a seismic shift toward sustainable practices, driven by increasing societal expectations around environmental stewardship and ethical labor practices. This has spurred the development and adoption of cleaner extraction methods, recycling initiatives, and even the rehabilitation of mined land.

Central to this transformation is the digital revolution, which has streamlined operational efficiency and opened doors to new possibilities. Advanced technologies such as artificial intelligence, machine learning, and virtual assistants are reshaping how mines operate, offering real-time analytics, predictive maintenance, and automated decision-making previously unthinkable.

Thus, the modern mining landscape is a complex interplay between tradition and innovation, where the past's tried-and-true methods coexist with the future's cutting-edge technologies. It's a fascinating ecosystem that continues to adapt and grow, constantly finding new ways to meet the challenges of resource extraction while minimizing its impact on both people and the planet.

### 5.1 Traditional methods and limitations

1. Open-Pit mining:

This is the most common method, where layers of earth are stripped away to reveal minerals or ores underneath. This method provides high production rates but at a significant environmental cost due to habitat destruction and topsoil removal.

2. Underground mining:

When deposits are deep within the earth, tunnels or shafts are dug to reach them. This method has a lesser environmental impact than open-pit mining but poses increased safety risks for workers due to potential tunnel collapses or hazardous gas exposures.

3. Placer mining and dredging:

Used for extracting minerals from alluvial deposits, placer mining involves sifting through materials in riverbeds or flood plains. This method can disrupt local ecosystems and affect water quality.

4. Mountaintop removal:

Predominantly used for coal mining in mountainous areas, this involves removing the summit of a mountain to access the deposits below. The resulting debris is often dumped into nearby valleys, significantly impacting the environment and local communities.

## 5.2 Limitations

- Environmental impact: Traditional methods, especially open-pit mining and mountaintop removal, can cause significant habitat destruction, water pollution, and air pollution.

- Safety concerns: Underground mining, in particular, poses various risks, including cave-ins, gas exposures, and long-term health effects like black lung disease.

- Operational efficiency: As easily accessible deposits diminish, more effort and resources are needed to extract the same amount of ore or minerals, leading to decreased operational efficiency.

## 5.3 The necessity for digital integration

1. Safety enhancements:

Digital tools can monitor environmental conditions in real-time, predicting and alerting of potential hazards. Automated equipment can also be deployed in dangerous situations, reducing human risk.

2. Operational efficiency:

Digital integration means better data. With sophisticated analytics, mining operations can pinpoint exactly where to mine, reducing the amount of wasted effort and resources. Predictive maintenance, enabled by digital monitoring, can foresee equipment failures before they happen, reducing downtime.

3. Environmental responsibility:

Digital tools allow for precision in mining operations. This means less unnecessary excavation, more accurate resource targeting, and minimized environmental disruption. Additionally, digital monitoring can track pollution and waste in real-time, ensuring compliance with environmental standards.

4. Stakeholder engagement:

In the age of information, communities and stakeholders demand transparency. Digital platforms can provide real-time data sharing about mining operations, environmental impact, and safety measures, fostering trust between mining companies and the communities in which they operate.

5. Future-proofing:

The mining industry must adapt as the global landscape changes through regulatory pressures, societal demands, or market shifts. Digital integration provides the flexibility and adaptability to navigate these changes, ensuring the industry's longevity and relevance.

*DOI: http://dx.doi.org/10.5772/intechopen.1003846*

In summary, the mining landscape, rich in tradition and history, necessitates a pivot from traditional methods to a more integrated, digital approach. This shift, while challenging, holds the promise of a safer, more efficient, and more environmentally responsible mining industry.

## 6. The role of virtual assistants in mining

The modern mining landscape is a testament to the harmonious integration of human expertise and technological prowess. Central to this transformation is the inclusion of virtual assistants, whose capabilities have revolutionized various facets of mining operations. Here, we dive into the multi-faceted role of virtual assistants within the mining sector, elucidating how they enhance efficiency, safety, and accountability.

### 6.1 Data gathering and analysis

1. Real-time monitoring:

Virtual assistants, in conjunction with Internet of Things (IoT) sensors, continuously collect data from different areas of the mining site. This includes metrics like temperature, humidity, equipment performance, and vibration levels.

2. Data integration:

These assistants can amalgamate data from various sources, ensuring a comprehensive overview. Whether it's data from underground sensors or satellite imagery, virtual assistants can consolidate and cohesively present them.

3. Advanced analytics:

Leveraging machine learning algorithms, virtual assistants can analyze vast data sets quickly, identifying patterns, anomalies, and trends. This enables mining operators to make informed decisions, from identifying the best drilling sites to optimizing transportation routes.

### 6.2 Predictive maintenance and equipment monitoring

1. Condition monitoring:

Virtual assistants keep a vigilant eye on the health of machinery. By analyzing data like temperature, pressure, and wear patterns, they can gauge the condition of equipment, ensuring that they operate within optimal parameters.

2. Proactive alerts:

Virtual assistants can send proactive alerts before a machine reaches a critical failure point. This allows for timely intervention, reducing unscheduled downtimes and expensive repairs.

3. Lifecycle analysis:

By constantly monitoring equipment performance, virtual assistants can provide insights into the machinery lifecycle, helping in planning replacements or upgrades, thus ensuring that operations remain efficient over time.

### 6.3 Safety protocols and accident prevention

1. Hazard detection:

Virtual assistants can detect potential hazards like gas leaks, high temperatures, or tunnel structural weaknesses using sensors and real-time data analysis. Immediate alerts can then be sent out to evacuate or secure affected areas.

2. Training modules:

Virtual assistants can utilize augmented reality (AR) to create immersive safety training modules. These interactive sessions can simulate potential dangers and teach miners how to respond effectively.

3. Emergency response:

In the unfortunate event of an accident, virtual assistants can quickly communicate with emergency response teams, providing them with crucial information like the location of thc accident, the number of people affected, and the nature of injuries, if any.

### 6.4 Automated reporting and documentation

1. Seamless documentation:

Virtual assistants can automatically document daily operations, from the amount of ore extracted to the hours logged by each machine, ensuring that records are accurate and up-to-date.

2. Regulatory compliance:

Stringent regulations often bind mining operations. Virtual assistants can generate compliance reports, ensuring operations adhere to local, national, and international standards.

3. Stakeholder communication:

Virtual assistants can create comprehensive reports for stakeholders, investors, community leaders, or environmental agencies. Automated yet personalized, these reports can provide insights into operational performance, environmental impact, and future projections.

In conclusion, the role of virtual assistants in mining transcends mere operational efficiency. They act as guardians, ensuring safety, as analysts, providing insights, and as communicators, bridging the gap between operations and stakeholders. Their versatility and adaptability make them invaluable assets destined to shape the future of mining.

## 7. Enhanced exploration with virtual assistants

Exploration, the fundamental starting point of any mining operation, exists in a unique realm where scientific inquiry's precision mitigates nature's unpredictability. Virtual assistants are invaluable allies in an era where the industry focuses on tapping into unexplored resources while balancing minimized risks and optimized returns. These advanced tools bring computational power to the table and introduce a level of analytical depth and agility previously unimaginable in the sector.

Gone are the days when exploration relied solely on geological surveys, rudimentary sampling, and human intuition. Virtual assistants, equipped with machine learning algorithms and artificial intelligence, can analyze vast data sets from multiple sources in real-time. This includes everything from geological formations, mineral compositions, and past drilling records to current market demands for specific resources. This integrated analysis can significantly reduce the probability of costly exploration errors, boosting mining operations' overall efficiency and financial viability.

Moreover, these virtual tools can assist in creating predictive models that offer foresight into potential environmental, logistical, or financial challenges. For instance, by identifying unfavorable geological conditions early on, companies can reroute their efforts to more promising areas, saving time and resources. Similarly, predicting equipment failures before they occur can facilitate timely maintenance, enhancing safety and productivity.

Virtual assistants also pave the way for remote exploration capabilities, offering the chance to investigate difficult-to-reach or hazardous locations without risking human lives. They can control drones for aerial surveys or guide robotic apparatus in underground environments, delivering a wealth of accurate data that can be synthesized into actionable insights.

In sum, the advent of virtual assistants in mining exploration is transformative, turning traditional methodologies into sophisticated, data-driven ventures. By doing so, these cutting-edge technologies are effectively revolutionizing how we discover and assess natural resources, setting the stage for a more sustainable, efficient, and lucrative future in the mining industry.

### 7.1 Geospatial analysis and predictive mapping

1. Data aggregation:

Virtual assistants integrate data from diverse sources like satellite imagery, ground surveys, and historical exploration data. The outcome is a rich and multi-dimensional geospatial database.

2. Terrain modeling:

Using advanced algorithms, virtual assistants can create 3D models of potential mining sites. This gives geologists and miners a virtual terrain representation, facilitating a detailed analysis before setting foot on the site.

3. Predictive mapping:

One of the standout capabilities is using machine learning to predict potential mineral-rich zones. By analyzing existing data patterns, virtual assistants can highlight regions more likely to yield valuable resources, streamlining the exploration process and reducing guesswork.

### 7.2 Risk assessment in new terrains

1. Environmental impact predictions:

Before any exploration begins, predicting its potential environmental ramifications is vital. Through their extensive data analysis capabilities, virtual assistants can forecast the likely ecological impact, from groundwater disturbances to habitat disruptions.

2. Geological hazards:

Unforeseen geological hazards, like land instability or underground gas pockets, can jeopardize exploration missions. Virtual assistants can predict these hazards based on historical data and current geospatial metrics, ensuring that miners are adequately prepared or can choose to avoid high-risk zones.

3. Cultural and social impacts:

Exploration, exploitation (mining), and mine closure are not isolated. It impacts local communities, sometimes even historically or culturally significant sites. Virtual assistants can map and juxtapose these areas against potential sites, ensuring that mining activities remain sensitive to local socio-cultural nuances. It is well known to us and recorded how the 'Gold Rush' had impacted human culture and society. Though the mining risk is significant, new terrains for future development are also not limited.

### 7.3 Optimizing drilling and exploration decisions

1. Precise drill point determination:

Drilling is both resource-intensive and expensive. Virtual assistants can analyze subsurface data and suggest the most promising drill points, ensuring that each drill yields maximum value and minimizing unnecessary ground disturbances.

2. Resource allocation:

Based on the exploration data, virtual assistants can recommend the optimal allocation of resources. Whether the number of drills, the machinery required, or the workforce allocation, these decisions are data-driven and tailored for efficiency.

3. Continuous learning:

As drilling and exploration proceed, virtual assistants continuously update their data models, learning from each new data point. This ensures that decisions become even more refined and accurate as the exploration process progresses, based on real-time feedback [5].

In essence, virtual assistants are transforming the exploration landscape in the mining sector. Converting raw data into actionable insights ensures that every exploration decision is informed and optimized for safety, efficiency, and sustainability. In an industry that hinges on uncovering the unknown, virtual assistants act as guiding lights, illuminating the path with the brilliance of data and analytics.

## 8. Training and skill development with virtual assistants in mining

Though steeped in tradition and often resistant to change, the mining industry is amid a technological metamorphosis. As digital tools become increasingly integral to operations—from data analytics to automated drilling—the corresponding shift in the required skill sets is inevitable and urgent. The workforce must evolve, and their training methodologies must adapt to this new landscape. Virtual assistants are pivotal in this transformative process, acting as more than just operational aids; they are evolving into dynamic educational platforms that impart new skills, refine existing ones, and ensure the workforce is adequately prepared for the challenges and opportunities of the digital age.

The role of virtual assistants in workforce training is multifaceted. These tools can offer personalized, on-the-job training modules that adapt to an individual's learning curve and job responsibilities. Whether teaching a novice how to operate new machinery safely or helping a seasoned miner interpret complex data analytics, virtual assistants can provide the correct information at the right time, enhancing efficiency and safety.

Additionally, virtual assistants can employ machine learning algorithms to analyze performance metrics and identify areas for further training or retraining. This proactive approach helps companies anticipate and address skill gaps before they become significant issues, thereby maintaining operational excellence and reducing downtime.

Virtual environments and simulation training, guided by virtual assistants, are becoming another invaluable resource. Workers can practice complex tasks or emergency procedures in a risk-free, simulated setting, gaining valuable experience without the associated real-world dangers. These virtual scenarios can be updated and altered to reflect the ever-changing conditions and challenges in the mining landscape, providing ongoing, relevant training.

By providing a flexible, real-time training environment that can be constantly updated and personalized, virtual assistants are helping to bridge the gap between traditional mining skills and the demands of a rapidly evolving technological landscape. These advanced tools are not just enablers of operational efficiency; they are becoming essential components in human resource development. In doing so, they ensure that the mining industry's most valuable asset—its workforce—is primed, skilled, and ready for the digital age [6].

### 8.1 Virtual reality (VR) and simulation in mining training

1. Immersive learning environments:

VR powered by virtual assistants can create highly realistic and immersive training environments. Whether it's simulating the deep confines of an underground mine or the vastness of an open-pit mine, trainees can experience the workspace without physical risks.

2. Hands-on training:

Miners can practice operating machinery, navigating hazardous zones, or managing emergencies through VR simulations. This hands-on approach ensures that they are better prepared for real-world scenarios.

3. Real-time feedback:

Virtual assistants can provide real-time feedback during training sessions. If a trainee makes an error during a simulated drill operation, the assistant can immediately point it out, suggest corrective measures, and sometimes even demonstrate the correct approach.

### 8.2 Continuous learning and virtual assistants

1. Adaptive training modules:

With their machine learning capabilities, virtual assistants can adapt training modules based on a trainee's performance. The assistant can introduce more intensive training if a miner struggles with a particular concept or operation.

2. Keeping up with technological advancements:

The mining industry's tools and techniques are constantly evolving. Virtual assistants update training modules regularly, equipping workers with the latest knowledge and skills.

3. Data-driven skill assessment:

By analyzing performance in training sessions, virtual assistants can identify areas of strength and weakness for every worker. This helps in tailoring individualized training programs and ensuring a well-rounded skill set.

### 8.3 Upskilling workers for a digital future

1. Digital literacy:

As mining operations become more digitalized, there's an imperative need for the workforce to be digitally literate. Virtual assistants can introduce workers to fundamental digital concepts, ensuring they are comfortable with new technologies [7].

2. Advanced training for specialized roles:

Not every role in the mining industry will require the same level of digital expertise. Virtual assistants can provide advanced training modules for positions that demand a deeper understanding of digital tools and analytics, ensuring that specialists are at the top of their game [8].

3. Preparing for a collaborative environment:

The future of mining will be a collaborative effort between humans and machines. Virtual assistants can train workers to interact efficiently with digital tools like drones, automated machinery, or data analytics platforms.

## 9. Summary

In conclusion, integrating virtual assistants into the digging for minerals industry development, including the training and skill development of miners as well as the mining workforce, the training and skill development of the mining workforce is not merely a progressive step but an imperative. As we witness an unprecedented convergence of the physical and digital worlds, the complexities that come with it require a workforce that is agile, versatile, and adept at navigating both realms. Virtual assistants serve as the bridge to this new reality, enabling real-time training, personalized skill development, and predictive analysis that helps pre-empt challenges before they escalate into problems.

These tools empower the industry's human element, enriching traditional skills with digital competencies. They elevate the workforce from mere operators to decision-makers who can leverage technology for improved efficiency, safety, and sustainability. This is crucial because, even as machines become more capable, humans remain indispensable for their creativity, adaptability, and ethical considerations.

Furthermore, as we move toward a future that demands increased sustainability and accountability, virtual assistants can train workers on best practices that minimize environmental impact and promote responsible mining. This ensures that the workforce is skilled and aligned with the broader social and environmental goals that are increasingly important today.

So, it's not just about meeting the challenges of tomorrow; it's about anticipating them and being prepared to adapt, innovate, and lead. In this context, virtual assistants are not just a technological add-on but a strategic asset integral to the long-term viability and competitiveness of the mining industry.

## Author details

Ali Soofastaei
Innovative AI, Brisbane, Australia

*Address all correspondence to: ali@soofastaei.net

## References

[1] Galar D, Kumar U. Advanced analytics for modern mining. In: Advanced Analytics in Mining Engineering: Leverage Advanced Analytics in Mining Industry to Make Better Business Decisions. London: Springer; 2022. pp. 23-54

[2] Khan MA et al. Correlative analysis and impact of intelligent virtual assistants on machine learning. In: 2019, the 11th International Conference on Computational Intelligence and Communication Networks (CICN). Honolulu, HI, USA: IEEE; 2019

[3] I Martin AV, Selva D. Daphne: A virtual assistant for designing earth observation distributed spacecraft missions. IEEE Journal of Selected Topics in Applied Earth Observations and Remote Sensing. 2019;**13**:30-48

[4] Alamdari S et al. Application of machine learning techniques to predict haul truck fuel consumption in open-pit mines. Journal of Mining and Environment. 2022;**13**(1):69-85

[5] Mariz JLV et al. Advanced analytics for rock blasting and explosives engineering in mining. Advanced Analytics in Mining Engineering: Leverage Advanced Analytics in Mining Industry to Make Better Business Decisions. 2022;**2**(1):363-477

[6] Martins P et al. Process analytics. In: Data Analytics Applied to the Mining Industry. London: CRC Press; 2020. pp. 131-148

[7] Soofastaei A, Fouladgar M. Advanced analytics for haul trucks energy-efficiency improvement in surface mines. Advanced Analytics in Mining Engineering: Leverage Advanced Analytics in Mining Industry to Make Better Business Decisions. 2022;**3**(1):539-556

[8] Soofastaei A, Knights P, Kizil M. Application of advanced data analytics to improve haul trucks energy efficiency in surface mines. Extracting Innovations: Mining, Energy, and Technological Change in the Digital Age. 2018;**2**(5):163-179